Saying No

Why It's Important for You and Your Child

ASHA PHILLIPS

First published in 1999
by Faber and Faber Limited
3 Queen Square London WC1N 3AU

This edition first published in 2008

Typeset by RefineCatch Limited, Bungay, Suffolk
Printed in England by CPI Bookmarque, Croydon CR40 4TD

A CIP record for this book
is available from the British Library

ISBN 978-0-571-24157-6

2 4 6 8 10 9 7 5 3 1

Contents

Preface to Second Edition

The response to *Saying No* on its publication in 1999 was enthusiastic (particularly in Italy, where it sold more than 600,000 copies). The book has since been translated into twelve languages. Many reviewers and parents expressed relief at the approach the book takes: that saying 'no' is beneficial rather than detrimental to children's well-being. The ideas articulated have stood the test of time and seem to cross cultural, religious and ethnic lines. If anything, the issues of establishing boundaries and setting limits in managing child behaviour have become even more pertinent since the book was first published. We now see frequent media coverage of toddlers being threatened with expulsion from nursery schools because of unmanageable behaviour; young children are reported to be acting like delinquent teenagers; and, in the UK, adolescents are facing Anti-Social Behaviour Orders (ASBOs).

Parents seem unable to cope with such behaviours, as the proliferation of television programmes on how best to deal with difficult children demonstrates. In the UK alone, we have titles such as *House of Tiny Tearaways, Who Rules the Roost, Driving Mum and Dad Mad, Brat Camp* and *Supernanny*, to name but a few, vying for space in the television schedules. All of them indicate a loss of confidence in parents' belief that they are capable of bringing up their children properly. Of central concern to parents is what limits should be set and how to set them. They worry about how much television very young children watch, or how much time they spend on the computer, be it playing games or on the internet. I hear from many who feel they cannot decide on appropriate time limits, nor can they monitor or enforce them. Faddy eating habits and difficulties in settling children to sleep are still prominent issues for parents in my client group. Given the overwhelming evidence of these common experiences, *Saying No* is more relevant than ever.

When I originally wrote the book, my aim was to be comprehensive, dealing with general questions and the uncertainties that ordinary families face. However, I have realized that there may be no such thing as an 'ordinary' family because so many face particular difficulties. In considering when it is particularly difficult to say 'no' and when it is especially testing to find appropriate limits, my practice and my environment have increasingly highlighted an often-ignored area. When there is illness in a family, be it physical or mental, ordinary boundaries get blurred. I have focused on the helpfulness of limits, yet it is clear that much more could be written about the impact of illness or disability on families. As this is too wide a subject for this book, I have added a final chapter which deals with the particular difficulties in saying 'no' in such circumstances. I illustrate my thinking with some examples from different age groups, following the structure of the original book.

I wish to thank my wonderful friend and colleague Janine Sternberg for her ever-helpful comments and advice on the new chapter. I particularly want to thank the families who have shared their journey with me. It is never easy, indeed it's often very painful, to articulate one's most personal and raw feelings. I am very grateful for the time they have given me, for reading the section relevant to them, and for offering such helpful comments and examples for me to include.

Asha Phillips
2008

Acknowledgements

Wording acknowledgements somehow makes this book seem more real to me than the hours spent writing.

First, at Faber, I wish to thank Matthew Evans for making me believe I could do this. I am especially grateful to Daphne Tagg, who with her sensitive copy-editing has put shape to what must at times have been rambling and meandering thoughts. My warmest thanks go to my editor Clare Reihill, who has been a source of encouragement and support throughout. She has looked after me and the production of this book every step of the way and I am greatly in her debt.

I am grateful to my students for allowing me to use material they have presented for supervision. I am very fortunate to have access through them to ongoing work in many different settings and the opportunity to keep learning.

I wish to thank all the children, families and couples I have seen, for sharing their struggles with me and for sustaining the hard joint venture of therapy which brings change.

Many friends and colleagues have lent their support, and unfortunately I cannot name them all. Sue Reid as my personal tutor and supervisor, many years ago, during my training at the Tavistock Clinic, encouraged me to find my own style and to trust my heart, as well as my mind, in my work with children. Her teaching was inspiring and I feel deeply grateful to her for that and her continuing friendship. Dr Claude Wedeles gave and taught me more than I can say; his insight, humour and affection as a psychoanalyst made me feel held and connected throughout my journey of discovery.

I am thankful to Dr Gill Stern who generouly provided case examples and to Gill Markless for her very helpful and clear comments. I wish to thank most particularly Janine Sternberg, who valiantly read through every single draft, gave telephone advice and always made time to respond to text faxed through at short notice. In our many conversations, she helped me formulate my thoughts.

I am also indebted to my parents, Jean and Freny Bhownagary, and my sister Janine Bharucha, for their critical reading, encouragement and suggestions.

I wish to thank my husband, Trevor, for urging me to think big and believing there was a book in an idea to which I might not have given a second thought.

My children have made the greatest contribution, as with them I have had to learn from experience. They have both been tremendously supportive, offering ideas and questions. They have also tolerated my preoccupation with the book with good humour and put up with readymade meals, snacks and sandwiches because I had no time to cook. Thank you!

A.P.

To My Children

Where the river of your life flows
By the banks of your green hours
I have stood still as a tree grows
Watchful over its spring flowers

Whatever else may seem lies
This magic has been granted me
To witness mirrored in new waters
The rising of suns
And moons drifting in the mystery of young eyes.

FRENY BHOWNAGARY

Introduction

> A repertoire might be more useful than a conviction, especially if one keeps in mind that there are many kinds of good life.
>
> ADAM PHILLIPS, *On Kissing, Tickling and Being Bored*

It seems so obvious that we should say no at times, yet the common assumption is that we should say yes whenever possible. There is an unspoken rule that if you are kind, caring, polite or thoughtful, you do not say no. This is all-pervasive in most walks of life, from the intimacy of our home to the public sphere of politics; even advertisements proclaim 'the Bank that likes to say yes'. In my clinical work as a child psychotherapist I often see families who have great difficulties in saying no, and this plays a major part in their predicament. I am also aware of how common this reluctance to say no is for myself and my friends.

I believe that saying yes whenever possible is not helpful. By not saying no at appropriate times, we may be robbing ourselves and our loved ones of capabilities and resources; we may be restricting ourselves, by not stretching our 'emotional muscles'. Saying no does not have to be a denial or crushing of another; in fact it can demonstrate a belief in the strength and abilities of others. It is the necessary corollary of saying yes; both are crucial. This book focuses on what it means to say no in the family context and why it is essential.

Depending on the philosophy of the time, children have been pictured in various lights, and the role of parents has shifted accordingly. Children have been seen as wild beings needing to be civilized, as in William Golding's *Lord of the Flies*, or as blank slates, material to be properly moulded. Alternatively, they have been represented as naturally good and likely to blossom if allowed to do so by simple encouragement and nurturing, in the image of the noble savage of the eighteenth-century French philosopher Jean Jacques Rousseau. We experienced a revival of this image in the 1960s with the mantra 'All you need is love'. These positions have recurred cyclically through time and appear with various prominence in different cultures.

Inevitably, approaches to child rearing reflect these fluctuating views of children. We have had very regimented, adult-led styles, which believe that the parents should organize and manage all their children's behaviour, such as patterns of sleep and feeding on a four-hourly basis for babies. We have also experienced the now common baby- or child-led ways. Following the 1960s this liberal style permeated all our approaches to children, at home and at school. We had feeding on demand for babies and very open-plan, child-centred education, exemplified at its most extreme by A. S. Neill's school Summerhill in Suffolk. What can be seen clearly is that, over time, our views and approaches to children have changed. At present, we live in a period which is not dominated by a specific approach. The generation of 'Spock' children whose parents felt they followed Dr Benjamin Spock's recommendations in the 1950s has no equivalent. Just as in music or fashion, there is no specific style for the current decade. This opens up a more creative space in which we can make up our own minds. It also leaves many parents feeling confused.

I believe that with many of our problems, we look for solutions that are familiar and close by. One of my favourite stories comes from the Sufis, who use jokes and tales about a figure called Mulla Nasrudin to illustrate philosophical points, and also for the sheer fun of them. The story goes like this:

> A man saw Nasrudin searching for something on the ground outside his house.
> 'What have you lost, Mulla?' he asked. 'My key,' said the Mulla. So they both went on their knees and looked for it.

After a time the other man asked: 'Where exactly did you drop it?'

'In my own house.'

'Then why are you looking here?'

'There is more light here than inside my own house.'

Like Mulla Nasrudin, we look where there is more light. Thus we go round and round, pondering repetitive, circular arguments and familiar thoughts, and get nowhere. I hope that this book will help the readers to venture back into their home and look where they might truly find their own key.

This is not a prescriptive book, nor a recipe book on how to say no. It aims to help you think about yourself and your family in relation to the ability to say no. I believe that if we understand our behaviour, as well as the impact it has on others, we have more choice in life. Inevitably a book of this kind tends to focus on our struggles and on how and why we make mistakes, but my emphasis is on processes, that is to say, development and change. With that in mind, there can be no universal solution to our dilemmas; what matters is finding our own means and a variety of them. As a psychotherapist, I strongly believe in growth and in the resilience of human beings. Often when we, as parents or teachers, read about problems faced by families, we worry that we are doing so much wrong and somehow ruining the lives of the children in our care. It is always important to remember that people are open to change and no difficulty need remain for ever.

The book is divided into chapters based on age. These are not separate, self-contained stages, but each chapter looks at what is predominant at a particular time. The principles applied in each section are valid for all ages. It is possible to refer just to the chapter which most concerns you, without going through the book. I do, however, recommend reading Chapter 1 on babies first because it explains much of the thinking which follows.

To protect confidentiality, names and details have been changed to disguise individuals and families, so that they will be unrecognizable, except possibly to themselves. The case studies are for illustrative purposes and are sometimes compounds of various people's stories. I hope that anybody who thinks they recognize themselves will feel I have been true to them and will not feel exposed. Whenever possible, and certainly with friends and students, I have received permission to use the material.

For convenience and ease of understanding I have referred to the child as 'he' and the parent as 'she'. In reality, the mother is the more frequent primary carer and most often child care outside the home is given by women. I hope that the importance of fathers, siblings and other significant people will be clear throughout the book.

1

Babies

'Where have I come from, where did you pick me up?' the baby asked its mother.

She answered half-crying, half-laughing, and clasping the baby to her breast, 'You were hidden in my heart as its desire, my darling.'

RABINDRANATH TAGORE, 'The Beginning'

Every child asks, 'Where did I come from?' We are all familiar with the facts of life, but we also know that our children are already present in our lives, before they are born. By this, I mean that we all have dreams of what a child of ours may be like, how we imagine him, what we wish for and what we fear. We wonder about what sort of parent we will be, we think of what kind of parent we do not want to be. All these thoughts and fantasies pre-date the baby and even the conception. The new baby is born to a couple (even if he lives with a single mother, father is present in him) and into a family but also into his parents' mental world, populated by figures from the past and present, hopes, expectations, dreads and a myriad other feelings. This first chapter hopes to address how parents and baby fit together within this constellation.

For a long time people thought that babies weren't 'proper' people, they were little creatures who just fed and slept and occasionally played for short periods of time. Modern research has shown that, in fact, new-borns and very young babies are tremendously sophisticated. Here are just a few of their abilities: they can see, smell, hear, taste; they can also discriminate and have preferences. We know that even in the delivery room, they prefer human faces to abstract patterns. New-borns can fix their gaze on you and imitate certain expressions, such as thrusting out the tongue.

They recognize the smell of their mother's milk and turn to it by seven days old. They prefer human, especially high, lilting, female voices to non-human sounds. They like sweet tastes and can recognize a change in the sweetness of liquids after only two sucks. All these skills are useful in holding their parents' attention and engaging their emotions. Many parents will think of this as something that they always knew or that their very special baby did. Conversely, they may have been told that their baby's smile was 'just wind' and so have disbelieved their own understanding of their baby's sociability. We now can show, reliably, that these social responses from infants are consistent and not just a fluke, later interpreted by parents as proof of the wonder of their offspring!

Research also demonstrates that infants are finely attuned to the behaviour and moods of those close to them. For instance a baby will often cry if the person holding him is talking about something very sad.

The other side to this sensitive, active and outward-looking nature is that babies can often feel flooded by stimuli. General noises, colours, parents and others looking right into their eyes, holding their hand, talking to them, can sometimes feel overwhelming and the baby may cry or withdraw (this is particularly the case with premature babies). We have all met those prickly little ones who react rather like hedgehogs, scrunch up their eyes, turn away and whimper when you are trying to be friendly. They may just be a bit over-stimulated and need you to tone down your enthusiasm.

Because babies are very complex individuals, what we do and how we approach them has enormous impact. The paediatrician and psychoanalyst D. W. Winnicott wrote that 'a baby cannot exist alone, but is essentially part of a relationship.' The growing body of infant research shows that what matters most is not what the parent or child brings to the encounter but what happens between them – the reciprocity of the interaction, the effect they have upon one another.

Interpreting Needs

In order for a person to develop and thrive, he has to feel loved and understood. Attunement to a baby's state and communication is crucial. When a baby expresses himself, through crying, wriggling, smiling, cooing, and a whole range of non-verbal cues, we interpret his wishes – for a cuddle, a feed, a change of nappy, some time playing or a host of other possibilities. The baby cannot be more specific and in most cases probably does not know what he wants; he is just letting us know how he is feeling. Responding to his individual communication and thinking about what has helped in the past gives the baby a sense of being held; it stops him from feeling he is falling apart. Have you noticed the movement that resembles an astronaut in a gravity-less space, arms and legs flying out to the side? This is how we sometimes see babies react when they do not feel held, as if they are somehow suspended in space with nothing to hold on to or ground them. Most parents' reaction is to gather the baby up in their arms and enclose him in them.

What we also do, perhaps less consciously, is gather the baby up mentally, translate his actions for him and give them meaning. Wilfrid Bion, a famous post-Kleinian psychoanalyst, wrote about the parent's role in helping to 'contain' their baby emotionally. He describes how, when an infant is bombarded by feelings which can be overwhelming, the role of the mother is to take those feelings into herself, digest and process them and then put them back to him in a more palatable form – she translates something that is unbearable into something which can be managed. For instance, the baby may be crying, with a pain in his tummy. The mother will calm him down by talking to him, explaining that it is just wind, hold him, pat his back, help him to burp. The baby, who may have felt he was collapsing, will then feel held, gathered together and later soothed and satisfied.

With this experience consistently repeated, he gets not only the security of being heard and made to feel better, but also a model of how somebody can deal with distress. It is the beginning of a model for thinking – thinking about feelings. Through this the baby learns to recognize and give shape to his own experiences, and begins the ongoing process of communication and mutual understanding.

Sometimes we get it right and sometimes we don't. Trial and error, attention and observation guide us. Imagine the following ordinary situations, all recorded from observations of different babies feeding. The extracts are quoted from students undertaking an Infant Observation course. This consists of making weekly visits to a family with a new baby over a period of one or two years. Each visit lasts an hour. The observer does not take notes but immerses himself in the atmosphere and later records in as much detail as possible everything that happened during his visit. He is not looking for specific data or trying to analyse what he sees, but simply observe what occurs. This naturalistic method is close to that used by ethologists when studying animal behaviour. A seminar then tries to understand the experience described. Jumping to premature conclusions and a judgmental stance are actively discouraged. The purpose is to learn about development through experience. Infant Observation is a key component of psychoanalytic training but is also very helpful to any professionals working with children or in a therapeutic context, such as nurses, health visitors, teachers, social workers or GPs.

Tim, six weeks old, woke up slowly, opening and closing his eyes a few times, while making soft complaining noises. His mother sat next to him and looked at him. She said in a childlike voice, 'Coming back from the dream world, are you? I wonder what you were dreaming. Did you hear the car hooting? What did you meet in that other world? Anything you need to tell Mummy?' Tim made a slightly louder sound, keeping his eyes focused on hers. His mother picked him up and held him against her. He stopped crying and made rooting movements. She said, 'You're hungry, aren't you? Want some milk in your tummy?' She un-buttoned her shirt and held him to the breast, he hesitated for a moment and then grabbed the nipple eagerly. Tim had his eyes focused on his mother's eyes while sucking and she talked to him gently, holding his gaze. 'You're hungry, aren't you? I wonder how the milk tastes in your mouth. It is sweet but you only know the sweet taste. I think it tastes different at different times, doesn't it?' Tim let go of the nipple, looked at her and then sucked again.

Here we see the mother very much putting herself in Tim's place, imagining what it's like to be him and trying to tailor what she offers to what she feels he needs. By her talking and looking she keeps in touch with him and puts words – a shape and a meaning – to his initial communication. His original complaint is taken notice of and she gives him the experience of being heard. She then translates his discomfort into a need for a feed, which she offers. At the same time, she is aware of his transition from sleep to wakefulness and treats him gently, helping him to wake up, soothing him with her voice and her thoughts. She is containing him.

Another example shows a more physical holding of the baby:

Hannah, also six weeks, lies in her mother's arms and becomes excited, anticipating what is to come. She turns her head and body in an active and purposeful search for the breast. Mother turns towards me and says, 'You see, she knows what is going to happen.' She cradles her infant's head in her arms, whilst at the same time presenting her breast, in such a way that Hannah latches on to the nipple straightaway: she does not suck for a second or two but waves her arms in the air and kicks, twitches a little as if to settle herself, before beginning to suck. Soon her body becomes still, her hands come to rest on the skin of her mother's breast. There is no further movement apart from her rhythmic sucking, punctuated by small pauses. I notice that her mother is holding her very closely, using both arms to enclose her. Their bodies seem to mould themselves into the shape of the other, so that the two appear as one. They are absorbed in each other. The room is silent for about twenty minutes. Hannah is asleep on the breast. I'm not sure then if the infant falls away from the nipple or the nipple falls away from the infant.

Here, the mother holds Hannah closely in her arms. The anticipation of the feed, which leads to Hannah being excited, is recognized by her mother, who offers her breast. Then, as her mother adds her own mood of calmness and enjoyment of the feed, they both settle. We can see the ease with which the moment can feel like a union, a two-in-one state of being; this permeates much early contact between mothers and babies but is all the more poignant for the fact that part of one's body is inside the other. In bottle feeding and later spoon feeding this is still present, as an extension of the self, but less graphically so.

Sometimes the feeding situation can get very stressful, perhaps as a result of a complicated birth, a mother feeling very lonely and isolated, or because she is just getting to grips with what this new baby means to her and her family.

Lucy's mother tells me, 'It was awful at the weekend. We went to see my parents. Lucy didn't settle at all. She kept on refusing the breast. I went to the clinic on Monday and discovered she had only put on one pound since she was born. It's so upsetting and unrewarding. I loved feeding Lucy's sister. Even the night feeds. Just you and the baby, all alone in the world, absorbed in one another. I'm not getting it this time.'

As her mother speaks, Lucy (aged seven weeks) is crying. her mother gathers a pillow and a magazine. She puts Lucy on the pillow. Lucy cries louder. Her mother could not remember which side to offer. She took several seconds deciding and I felt afraid she would leave it too long. Once the breast was offered, Lucy took it immediately and began to suck quite steadily, fingering the breast, her feet moving slightly. Then her legs convulsed and she let go of the nipple and gave a cry. Then she turned back, crying slightly and sucked again, twisting her body. Her mother took her off the breast and said, 'Let's wind you a bit.' Lucy, sitting on her knee, bellowed furiously, face contorted. Her mother rubbed her back and tried to burp her, without any result, while Lucy cried. Her mother put her back to the breast and once again Lucy took the nipple and sucked for about thirty seconds. She passed wind and began to cry, her body curling up. Once again her mother sat her up. Lucy's face became dark red as she screamed. This pattern was repeated during the entire feed.

Here we see a mother already anxious about the coming interaction; she describes how upsetting she finds the feeds; she prepares herself for this, getting a magazine to take her mind off it. The interchange is permeated by uncertainty and difficulty in matching and coming together. This would be what Daniel Stern, an American professor of psychiatry and researcher, has called 'missteps in the dance' in his book *The First Relationship*. The experience is not satisfying for either party; it leaves them both uncontained and separate and the mother feels she is failing.

Interaction: Matching and Mismatching

In most ordinary families there will be times of matching and mismatching, of being in harmony and of being out of sync. These are the everyday hiccups of growing up. We hope that, on balance, the positive predominates and helps the child and parent overcome the upset or disappointment of the bad times.

By attending to and interpreting the baby's communications, the parents help him to become settled and involved in the world. An infant benefits from a sustained experience of parents fitting in with his needs and accommodating him. Without this solid base, he will not be able to bear frustrations and waiting.

To explore an infant's sensitivity to adult communication, experiments have been carried out which disrupt the interaction between mother and child. Studies have shown that a baby's interaction with a parent is characterized by attention and little breaks, a rhythmic pattern which allows both participants to take turns and to affect the pace. Each one's reaction dictates the other's response. Early studies presented fussing babies aged three weeks with a nodding, silent face. The babies cried and then stopped when the face withdrew. Well documented research which has been undertaken for some years involves 'still face' studies. The basic idea is that mother and baby are placed in a room and both are videotaped. The baby is seated in a comfy chair. At first the mother is asked to interact normally with her baby but not to pick him up. Then she withdraws from the room for a short period. When she returns she is asked to present a still, expressionless face to the baby for forty-five seconds. Both his and her reactions are monitored and video-recorded. The results show that a non-responsive face distresses and inhibits the baby. The effect is dramatic: the infant almost immediately detects a change and tries to elicit a response. He will usually then turn away and then back to his mother again. He may try this several times. When he has repeatedly failed, he will slump, become withdrawn and try to comfort himself. The videos also show that seeing the baby's response upsets the mother, who becomes agitated and withdrawn herself.

Lynne Murray, a British researcher, devised other experiments with video recordings of mothers' and babies' interaction. Initially, a mother and baby are filmed interacting normally. Then the tape showing the mother is played to the infant some thirty seconds later. The baby sees his lively and attentive mother, but as it is a replay and not live reaction, she is no longer tuned in to his specific behaviour. The baby then becomes confused and distressed. Lynne Murray made the distinction between 'naturalistic and non-naturalistic perturbations', showing that ordinary disruptions, like speaking to somebody else, made the infant quieten, but 'incomprehensible' behaviour from the mother, such as a blank face or mis-timed responses (smiling cheerfully when he is protesting), led to the baby becoming at first puzzled, then distressed and confused. In these experiments the mothers, too, became distressed when shown video recordings of their babies' reactions.

The experiments may sound rather cruel but it must be remembered that these are isolated moments lasting only a few minutes in a baby's life. However, the implications are clear for what would occur if this were the consistent experience between the parent and the baby. Lynne Murray's work has led to further research on families where the mother suffers from post-natal depression. This research has been very useful to detect early signs of disturbed interaction and so provide intervention and help.

These experiments and others show that it is the interaction between parents and babies which matters – the specific responses to the individuals involved. Babies need to be seen, heard and responded to in order to thrive. However, research also shows that a mid-range of responsiveness works best for healthy development – that is, when a parent makes errors in interpretation and the parent-baby couple recover. It is reassuring to know that as parents we are not expected to 'get it right' every time. It is also very pertinent to the thesis of this book that recovery from a mismatch promotes development, assuming that the baby's needs will be met to a greater rather than a lesser extent. Again to quote Winnicott, babies need 'good enough mothering'. The ability to say no must be accompanied by a sensitivity to a baby's needs. One of the fine lines to be drawn is when it is OK to start saying no.

I am reminded of baby Jim, whom I observed for the first two years of his life. His mother was very attentive and thoughtful, she seemed always to know what he wanted, and often anticipated his every need. At the time I thought she must surely be the ideal mother. When Jim was eleven months old and not yet walking, he loved to hold on to his mother and with her help 'climb' up and down the stairs. She would hold his hands and he would launch himself up with no regard for his mother bending over to support him. He would demand this activity for long periods of time, and she seemed unable to make a stand about when to stop. She became exhausted, he grew ruthless and tyrannical. I had to think again and understand that there is no such thing as an ideal mother.

What seems like a perfect fit – a mother who spares her child irritation of any kind – does not help. Over time, I realized that Jim's tolerance for frustration was very low and that he really found it hard to manage difficulties. His mother indulging him did not build up his strength, neither physically, as he was not using his own muscles to get up the stairs, nor emotionally. He also seemed to believe that he was doing it all by himself. He was thus denied the experience of developing his own means and the understanding that he needed her help. He was therefore unable to ask for assistance. Instead he would bellow a demand, or rather an urge, which then was satisfied. He may have felt that it was the strength of his will to get upstairs that got him there. By not recognizing his mother's role, he was unable to develop a sense of gratitude.

By saying no, his mother would have given Jim an idea of what he could or could not do by himself, as well as of what she could manage easily and what this activity cost her. Her reluctance to stand up to him encouraged him to become a little despot. This then percolated into their relationship generally, and time together was often unhappy – the mother felt bullied and helpless, while Jim was cross and demanding.

For me, this observation highlighted the fact that what is appropriate at one age may not be so at another. Babies and young children move and learn at a very fast pace; we need to adapt to their shifting needs. In this example, Jim really benefited from his mother's immediate responses when he was very little; it gave him confidence and a sense of being held and loved. However, later it stifled the growth of his independence and fostered his sense of omnipotence. It was as if it was impossible for them both to tolerate the in-between stage – that of not managing by yourself but trying and slowly achieving. The options were success or failure, not the necessary learning on the way.

In order to learn, we need first to be in a position of not knowing. If you think you know everything already, you cannot listen to anything new. To become stronger, you have to recognize that you can't do everything immediately. To gain from others, you have to feel they have something to offer. Being in touch with being dependent is the first step in asking for help and then making good use of it. The foundations are laid in infancy. Many young children and even adolescents live in a state of false independence and pseudo-maturity and find it very difficult to learn from teachers or benefit from good care.

Separate Beings

Saying no communicates that you are a separate being. The beginnings of being on your own, of separateness, are very important. In the early days a baby's capacity to manage on his own is very limited. Unlike other mammals, the human infant is extremely dependent for a very long time. With a parent who responds quickly to any cry or communication, the baby may well believe that he is not separate at all. He feels alarmed, he calls out and before he knows what is what, there is his father's or mother's face smiling at him over the cot. If this happens every time, the baby may have no feeling that the parents have a life of their own. Winnicott writes, 'exact adaptation resembles magic and the object that behaves perfectly becomes no better than an hallucination.' If there is a gap, the waiting establishes in the baby's mind the reality of the person who comes to him when he calls.

There is no such thing as a perfect parent. The idea that one could see to the child's every need and spare him every pain would in fact lead to an unhappy and mal-adapted child. It would not prepare him for life in a world inhabited by others; initially it would be a magic kingdom where he is king, but over time it would turn into a very lonely and unreal place. Think of the story of Prince Siddhartha, whose parents wanted to spare him any sight of ugliness or unhappiness. They kept him in their beautiful palace. In spite of their power and riches, they were unable to keep him sheltered. He eventually went out into the world, discovered the suffering of others and became the Buddha. Many other myths and folk tales illustrate the point that all the riches in the world cannot replace true human contact with others, even if this involves sorrow and pain. Real connection involves frustration, struggle and hate as well as comfort, harmony and love.

Response to a baby's communication gives him the feeling he exists and is real. A little space between his communication and a response starts to give him an idea that he is part of a greater world. The length of time spent waiting is where judgement comes in. As we saw with the studies mentioned earlier, an absence of response or an inappropriate response can be very alarming to the baby. How can we get this right enough of the time?

Very often a baby just starts to make a sound and someone picks him up and takes action – a change of nappy, a feed, the offer of a toy. In trying to be the perfect parents, whose child never feels frustrated, we sometimes interpret too early, before he has had time to taste his own feeling. We attribute meaning to the beginning of an urge and can then deprive the baby of actually having the full feeling. By wishing to spare the child, we may in fact rob him of his own experience. We may make him fit in with what comforts us when we are uneasy. We may be unable to bear the wait, which would allow the baby and us to clarify what would be most beneficial. We would in effect be unable to say 'no' either to the baby's call or to our interpretation of it.

Instant Comfort

The idea of a gap between a whimper and a response is crucial to development. Let us look at some situations where it is important to be able to say no.

SLEEP

James was born following a rather prolonged and difficult delivery. His mother, Ellen, is tired and often tearful for the first two or three weeks. James is very alert and responsive to his parents, spending many hours awake, snuggling into his mother, looking up at her adoringly, and generally being a delight. However, he hates being put in his Moses basket and whimpers as soon as he is out of contact with his mum. He does not sleep much and therefore neither does Ellen. As they both get more exhausted, they become more irritable. What began as pleasurable contact turns, for Ellen, into an inability to part. Ellen ends up carrying him in a baby sling all day long, as she cooks, cleans and generally gets on with her chores. She feels she cannot even go to the toilet on her own. James seems stuck to her, even occasionally like a parasite, living off her rather than relating to her. She feels this has nothing to do with who she is, as an individual, and finds him most irritating at times.

This is a common scenario, where mothers talk of babies becoming spoilt and 'knowing' how to get to them. A cycle of unhappy feelings can become a pattern of relating, for a while, in which neither partner is fulfilling his need. At times like these, as with Ellen, you will often find a mother walking up and down with her baby, asking him what the matter is; what does he want? Why isn't he happy?

In the case of Ellen and James, when the father, Nick, came home, he was able to take James and to see that actually James was exhausted. With Nick's help, Ellen was able to put James in his basket for a nap, in spite of his protestations, and to let him cry for a little while, after which he fell into a deep sleep and later woke, refreshed.

There are many important lessons here. Ellen needed someone outside herself and James to make sense of what was going on. She was too caught up with James to be able to separate herself. The supportive role of fathers, or grandmothers, aunts, good friends, is invaluable at this time.

We have seen the story unfold from the adult perspective; let us now try to imagine how it may have felt to James. He is used to skin to skin contact with his mother, he likes her smell and the feel of her clothes. He finds that his body just moulds into the curves of her arms, which accommodate the changes in his movements. She responds to his coos and expressions. How different the unyielding basket, with crisp sheets smelling of washing powder! Even when he wriggles to find the side, it is not as comforting as those arms. He may well feel quite bereft without her, as if he cannot manage in the basket. By always carrying him, or picking him up as soon as he whimpers, Ellen is reinforcing the feeling that only she will do, that truly the basket is an awful place to be.

On the other hand, by placing him in the basket and soothing him into it, by talking and settling him in, she is showing him it is safe and good for him to sleep in. She is saying no to his wish to remain in her arms and stating that at this point he needs to sleep and this is the best place to do it. By allowing him to grizzle and cry, she is hearing his complaint but she believes he will survive and holds on to what she knows he needs: rest, which he gets less of in her arms. By doing this consistently, she builds up a picture for him that he will be all right and so strengthens his sense of self. She also gives him the time to find his own means of adapting. In fact, after a while James discovered that if he wriggled, he could find a comfy niche against a little rolled up towel that Ellen put in his basket for him.

Other babies find their fingers, thumbs or particular positions which comfort them. Some will look for a bodily comfort, a soft or sweet-smelling object, some may prefer sound, and others may use their eyes to fix on a light, a plant or a pattern on the side of the cot. These choices may come from within and a sensitive parent might observe this, or it may be something which the parent offers as comfort, such as a favourite piece of music. A routine of cuddle, music, a few pats on the back and a soft goodbye may become a recognizable pattern for the baby, who will over time become used to the beginning, middle and end of the sequence. This then becomes familiar and reliable. The baby also learns that at the end of the nap, Mum or Dad will be there. All these are focal points which help him to settle. They may apply not only to sleep, but to any time that the baby is required to spend on his own – in a bouncy chair, on a play mat or wherever.

This is the beginning of an emotional stretch – an ability to reach inwards and develop resources rather than relying totally on the outside world to provide. Again, this capacity depends in great part on the baby being given enough love and understanding – the 'emotional fuel' which allows him to travel a little way on his own.

An immediate response may rob the baby from developing a sense of himself, by himself, which could give him great pleasure.

Michael, nine weeks old, was stirring in his sleep. He frowned slightly and then started to smile. The cradle he was in rocked gently with his movement, as it was suspended by hooks from a frame. For the next half hour, I was fascinated by his activity. His face changed expression constantly, seemingly without effort or distress. He moved one hand around over his face and mouth. He pursed his mouth, sometimes blowing small bubbles. His eyelids fluttered occasionally as though he was trying to open them, but he seemed just below the level of consciousness and would subside back into sleep. His legs moved, his movements became stronger and suddenly his eyes were open. He made a slight mewing sound but did not cry. He was looking up at the scalloped edge of the curtain, over the crib, which was waving slightly with his movements. His hands waved in the air and his eyes moved from right to left, looking at both the waving edges. He started kicking his feet excitedly and the cradle rocked under him as the curtains flapped. He had freed both his hands from the quilt and waved both his fists in the air as he kicked. His eyes were wide open, engrossed in the waving fabric above him, and his mouth formed a wondering O shape. Now and again he made delighted little noises. He seemed really pleased and excited.

If Michael had been picked up as soon as he originally stirred and started to emerge from his sleep, he would have missed a chance to explore his hands, to taste the bubbles in his mouth, to examine the movement of the curtains, to notice the impact of his movement on the rocking of the cradle – to get a sense of himself on his own, free to take in the world at his own pace. This opportunity gave him not just a moment of pleasure, but also the emotional experience of waking gently alone, of learning that he can explore his surroundings and make something of this experience by himself. He is starting to use resources within him rather than relying on an outside presence to hold him together or interest him. This possibility can arise at many other times, not just upon waking, whenever a baby is on his own. This brief moment carries in it the seeds of independence and self-confidence.

FOOD

Sometimes when we reach for quick solutions to a baby's discomfort, we do what seems right to us, not necessarily what is right for him. Health visitors frequently see parents who immediately offer food as comfort, when at times something else may do, such as talking or singing to the baby, holding him in your arms or even just with your gaze. If food is always the first solution, the baby learns from repeated experience that only food will do when he is complaining. Here are some examples from Infant Observation.

Julie, twelve weeks old, has been lying on her back, amusing herself for a good ten minutes with a play centre standing over her. Her mother, Paula, comes to chat to her and strokes her cheek and then leaves to do a few chores. For a minute or two Julie resumes kicking, gurgling and staring behind her. I was surprised, as I thought she'd object to her mother leaving. Then her gurgle turned to a more urgent sound, not yet a cry. Then very quickly, though not totally suddenly, she began to peel her lips back, reveal her gums and the moan transformed into a slow cry. Her face changed colour to a bit red, her eyes shut and her fists clenched. Paula came in and said, 'Poor Julie! Are you hungry?' and put her finger in Julie's mouth. Julie did not suck it as she has done in previous weeks. Paula said, 'You're hungry. Hang on, hang on.' She picked Julie up and she quietened.

Paula tried to recall which breast to feed her from and Julie looked at her. She settled for the right breast, Julie latched on and her mother said, 'good girl, good girl.' Julie sucked quietly for about eight minutes, sucking, pausing and sucking again. After a while Paula took her off the breast and said, 'Is that enough?' Julie looked bamboozled, a bit wobbly and uncomfortable. I was reminded of Roman emperors who gorged and then vomited.

Paula left the room to answer the phone and Julie looked a bit bereft. She began to pucker and cry. Mum returned and said, 'Oh, you're still hungry, sorry.' She fed Julie on the other breast. This time she latched on hungrily. Quickly she got into a rhythm of sucking and stopping. The feed lasted another fifteen minutes or so. Her mother said, 'I'm going to take you off for a minute, baby,' and made a phone call arranging lunch for that day with a friend. They discussed what they might eat. As Julie grumbled, Paula cut short the call saying she had to go and finish Julie's feed. Julie sucked some more. After a few minutes her mother removed the breast again. Julie now looked really floppy and shut her eyes. She lay asleep, looking big, satiated and well. Paula made another phone call to a friend to arrange lunch for another day. Again they discussed the menu. She then went to fix herself some lunch and returned looking more buoyant. She stared at the sleeping baby adoringly and stroked her tummy saying, 'good girl.' Julie barely stirred, her lips occasionally went into a sucking motion, as if remembering the feed.

We can see from this example that the mother immediately understands Julie's cry as one of hunger. It could, in fact, have had many interpretations: maybe Julie wanted some company, or a change of scene; maybe she was uncomfortable, having been on her back for a while and unable to change positions by herself. Julie starts by feeding quietly and at the first break looks totally filled up, as if she had fed when not really hungry. However, as the feed proceeds with various interruptions, she seems to feed more energetically. We could assume that she was hungry after all, or that she is learning to fit in with her mother's interpretation of what makes you feel better – a feed. She is also learning that her mother praises her for feeding consistently as at the end she strokes her rounded tummy.

From the mother's behaviour, we can see that she herself needs a feed, contact with her friends on the telephone, and also the anticipation of eating various meals. Eating is a common source of comfort for all of us, children and adults alike. In this example, it seems Julie's mother reaches out for what she knows makes her feel better and offers it to her baby. This is a very common reaction. Another option would have been to wait and see if Julie, by her behaviour, could indicate what she needed.

As this particular observation proceeded over the weeks, a

pattern of feeding Julie at every cry, even a discreet one, became established, and Julie's main contact with her mother was at feeds. There was a lack of other forms of interaction, such as Julie simply being held and chatted to or shown toys. Over this period Paula appeared rather unsupported, a little depressed, very much in need of company and actual food. As she became more relaxed and at ease in her new role of motherhood, she was able to find other activities and interactions which pleased Julie and herself. By the time Julie was five-and-a-half months old, she spoke lovingly to her, saying, 'There's more interesting things in the world than food, isn't there?'

Much of the focus on food may come from a mother's insecurity that she has nothing else to offer. At least the milk comes from inside her and is seen to benefit the baby. Many mothers talk of the baby not being interested in them as people, only wanting the milk. The feed then starts to represent what the mother feels is her only contribution.

Mothers who chose to bottle feed may also feel that is how they can offer their baby the best. They may not believe that what comes from them is any good and overcompensate by giving too much formula milk.

Anna had finally managed to have baby Carl after two miscarriages and a very difficult delivery. She felt terribly unsure about her capacity to care for and keep Carl alive. The early contact was hesitant, with Anna feeling she did not understand his signals.

Anna picks up a crying Carl (aged three weeks) and puts him on her shoulder and pats him. Carl tries to suck his mother's skin. Anna sits him on her lap and then places him back on her shoulder, where the baby starts to suck her collar. Anna says, 'I don't know what he wants, what haven't I tried? I know! I haven't put him on his tummy.' She lays Carl on his stomach, across her lap and strokes his back. Carl tries to raise his head and cries. Finally Anna tries a feed. She puts him to her breast, which Carl takes straightaway. His body and legs relax, curved around the contours of Anna's body.

Unlike Julie's mother, Anna's first thought is not that her milk is what Carl wants. She thinks of many other possibilities before she imagines that Carl may be hungry or simply want to suck. Carl put on weight, feeds well and happily, but over the next few weeks Anna believes she has not got much milk, and changes to formula in a bottle. Carl keeps gaining weight and reaches the top of the scale for children his age. A pattern gets established where Carl seems to want to feed longer than Anna wants him to and Anna describes him as feeding constantly.

Carl (aged eleven weeks) is crying and putting his fist in his mouth and Anna says, 'Surely you cannot be hungry again, you have just been fed an hour ago! Let's try you with the end of that feed.' Anna gives Carl the bottle with a few ounces left. When the teat is in his mouth. Carl looks startled, excited and relieved all at once, and sucks vigorously. He finishes the bottle quickly and starts crying again. Anna gets another bottle and Carl has exactly the same response. By the time he finishes, there is milk running down his face and he looks stuffed.

By this time, Carl seems to want to feed most of the time. It is hard for Anna to say no to him, to know when he has had enough. This is not an easy task, even for us as adults; when is enough ever enough? Anna and Carl got into a mode of relating based on feeding. Anna felt Carl wanted it and Carl could not stop. Neither of them could say no to this cycle of constant feeding. It became difficult to introduce other forms of being together, of Anna offering a different kind of 'feed', one that might involve conversation, interaction, giving Carl food for thought, other experiences to digest, rather than actual physical nourishment. Patterns are hard to break and changes within them difficult to perceive. When Carl eventually gave new signals, it took time for Anna to recognize them.

Carl (aged twenty-one weeks) is lying on the floor, interested in the toys around him. Anna picks him up and Carl gives a couple of whines. Anna says, 'Oh, you're hungry, I'll get you some milk.' She puts Carl back on the floor and goes to fetch the milk. By the time she returns Carl seems content, looking at a toy dog. Anna picks him up and starts feeding him the full bottle of milk, which Carl drinks.

Here we see how difficult it is for Anna to notice the change in Carl, his interest in the toys. The habit of feeding takes over and they both return to the familiar. Other babies might protest and turn their face away, or be clearer about their wish to play, by leaning over or stretching towards the toy.

In this mother-baby couple, there was a fit which was not helpful at the beginning. Over-feeding Carl, not saying no, led to him never being satisfied, always hungry or needy and becoming overweight. Over time, Anna was able to make more of a stand, to offer Carl water and eventually finger foods. Anna's anxiety following her previous miscarriages, and Carl's own difficult birth, made it hard for her to see that he was putting on weight and thriving. Her first insecurity about whether breast milk was good enough led her to a shift to formula. Even then, it was hard for her to recognize that he was healthy and well. Only after the worry about ensuring that Carl would survive had diminished was Anna more able to establish ordinary boundaries and to see other ways of being with her baby.

We have seen that at the early stages of coming together, it is quite difficult to understand the baby. His communications are interpreted according to our own reactions to feelings of neediness and our ideas of what nurture means. What we make of our baby's communication helps him make sense of his feelings too. With Julie and Carl, in their different ways, their need was interpreted by their mothers mainly as hunger and so they learned to feed and in Carl's case to demand food whenever he felt a gap, a feeling of neediness, inside him. These patterns, if unchanged, can lead to problems in later life, but with most families they are transient and soon overcome, as they were in both the examples above. They are the ordinary mismatchings of two different people coming together at a sensitive and vulnerable time for both.

ACTION AND EXCITEMENT

The physical aspects of caring for your baby – the feeds, nappy changes, baths, naps – are clear needs which most parents acknowledge. The more discreet emotional care, talking or singing to your baby, the closeness of holding him and engaging him in interaction, play an equally important part in his well being.

Some mothers find it hard just to be with the baby; they feel he needs to be entertained all the time.

Rosie, four-and-a-half months old, is sitting in her baby walker playing with a large plastic toy with buttons. As each button is pressed, a sound comes out of a speaker at the side. She is animated and concentrates hard. She looks at the buttons with wide eyes and pursed lips, then opens her mouth wide and dribbles slightly. Her mother, Liz, joins in, asking about the sounds and flicking a switch which alters them. Rosie looks at me and Liz and returns to her play. After a while of wriggling and banging, her position shifts so that she is leaning uncomfortably in the walker. Rosie gets frustrated and bangs harder and tries to get parts of the toy into her mouth. She starts to grizzle. Her mother takes the toy away and offers her a cloth book. Rosie puts it in her mouth and bites it. She grimaces and takes it out and looks at it. Liz then offers her a teething ring, which Rosie bites on firmly. She chews on it hard, flinches and starts to cry. Liz picks her up, Rosie arches her back and begins to slide into a lying position. Liz stands her up again, trying to make her bounce on her legs. Rosie protests and Liz lies her down and plays an exciting game with Rosie's feet, lifting them up and down. Rosie looks at her mum and pushes her tongue in and out and rubs her lips together. Liz then plays pat-a-cake with Rosie, who blinks a lot, especially at the end of the rhyme, when Liz finishes the game by touching Rosie's nose ('for *Rosie* and me!')9

The doorbell rings and a visitor arrives. The friend quickly reaches out and picks Rosie up, who looks uncertain and serious. The friend chats and jigs Rosie up and down a little trying to elicit a smile. Rosie holds herself stiffly, she becomes restless, wriggles, arches her back and starts to cry. Liz picks her up and holds her close. Immediately Rosie relaxes into the shape of her mother's body and snuggles up to her. Gradually her eyes begin to close and her breathing becomes shallow as she falls asleep.

At the beginning, we can see a baby who is very able to entertain herself with toys. However, as she becomes more uncomfortable and tired, her mother responds by offering more and more toys or activity to entertain her. Eventually, even the friend tries to jolly Rosie along. It is only at the point of the final cuddle that one feels Rosie gets what she may have wanted for a while. Until then, activity, change and excitement are offered as relief. Mum tries different sorts of stimulation as solutions, rather than waiting and observing Rosie to get an idea of what she may be saying.

This is a very ordinary example. Most people feel helpless in the face of a cry or a complaint and want to 'make it better' straightaway. We often feel we can achieve this by 'doing' something.

When the response to discomfort is always action, a baby learns that only activity makes you feel better. What does this dynamic mean to the mother, and to the baby? What message is being communicated? It could be that a mother is saying, 'I cannot bear to hear you cry, let's quickly do something to stop it.' At this point the mother's discomfort may be or become as great as the baby's. This will then increase rather than decrease the baby's distress. Instead of having a person to calm and soothe the cry or to give it a shape, the baby may be filled by his mother's worry as well as his own. A helpful response to the baby's cry could be, 'Don't worry, things are fine, you are just tired (want a cuddle, a feed, a little chat).' This would help baby and you to have time to see what he might really be feeling, or for you to help him find a way to manage his discomfort.

By choosing to be very active in your response, you could also be saying, 'Here, let's take your mind off what's bothering you.' A distraction comes in handy from time to time, but when it is always used as a method of dealing with upset, you are also communicating something else. You are saying, in an indirect way, that complaining is at best not acceptable to you or at worst unbearable. From the baby's point of view, what may start off as a little grizzle, a whinge that just needs to be expressed, may be turned into something that needs to be dealt with. If the complaint is cut short, the baby may soon learn that such feelings are unacceptable and he will need to find ways of dealing with that himself. Sometimes we are all a bit grumpy; it should be acceptable to have a moan and a groan without feeling this is not tolerable. Parents may need to learn just to stay with the grumpy baby, accept his complaint and offer sympathy: 'Yes, I know you're feeling miserable, we all feel that way sometimes, it's OK, things will improve . . .'

By managing a baby's grumpiness, the mother is not just helping him to overcome that particular moment, but is giving him a model of how to deal with difficulties. By tolerating his discomfort, his mother is then saying that this is an acceptable and bearable feeling and that they will both feel a little awful but, in the end, it will be all right. She is reinforcing the idea that being upset does not mean the end of the world, but is an ordinary pain that can be overcome. This helps the baby build a picture of himself and the world as safe. Learning to survive problems helps enormously in building up resilience in oneself and faith in others.

If, from very early on, activity is offered as the principal way of dealing with discomfort, the baby himself will adopt that model, as we see in the following situations.

Leo, ten months old, has just mastered walking around, holding on to furniture. He is quite secure on his feet but needs supervision. He enjoys racing around the lounge in this way, calling out to his mother when he needs help to cross from one chair to another. He does not want to crawl the space in between. He shrieks with excitement as he tears around the room and yells at the point when he wants his mother to assist him. She helps him and they spend extended periods of time like this. The activity gets boring for her, but she cannot bear his cry when she does not comply and worries at the over-excited state he gets into. Leo will not stay still for a minute and even sitting for meals is becoming awkward.

Celia, seven months old, has many toys and sits like a little queen on a mat surrounded by them. She is not very mobile and points and grunts in order to get her mother to provide the toy she wants. Her mother spends hours a day playing with her. Celia demands one game after another, making clear with her gestures and sounds what she is after. However, if her mother needs to make a phone call or cook a meal, Celia will throw herself down and scream. As soon as her mother returns she is happy to play quietly again. Her mother does not know what to do. She is pleased Celia can concentrate for so long and be entertained but resents having to be there all the time. She also worries that Celia is only happy when kept busy.

When activity is the regular method of dealing with upset, the baby often becomes over-excited, the mother gets exhausted, may feel taken advantage of and abused, or almost claustrophobic, wanting to get away from him. The baby in turn becomes terribly dependent on exciting interaction and may find it hard to be on his own. He becomes easily distressed if left to his own devices and cannot find a way to entertain himself. This has a flavour of the baby being addicted to his mother or the activity, rather than helpfully dependent. He acts as if he cannot manage, would collapse without her. They get tied up in a cycle which is overcharged with emotion and leaves them both feeling unhappy. As we saw with food, the feeling of an empty gap becomes intolerable, for both parent and child. This leads to compulsive behaviours and restricts openness to new things. If you instantly fill a gap, it is usually with something familiar. It does not permit creativity or the emergence of the new. It also makes for a rather emotionally charged atmosphere where calm is hard to establish.

This can have repercussions later on in the 'I'm bored' syndrome of primary-age children which drives parents nuts! And it may set a pattern for the future where it is very difficult for the child to linger, to ponder, to explore, just to be thoughtful and attentive, instead of restlessly reaching out for action. This can have an important and inhibiting impact on the capacity to play and to learn.

As the baby grows more able he will get an enormous sense of satisfaction from achieving goals by himself. We have all seen the excitement a baby expresses when he bangs two hard objects and a big sound is made, or he stretches to reach a toy and actually manages to hold it. I remember with admiration how my daughter spent extended periods of time pulling herself up to standing, flopping down and pulling herself up again. It was a delight to witness the pleasure she got from mastering her body. There are numerous instances of young children trying and succeeding for themselves, which go to building up their self-esteem. If they do not have to struggle at all, they may never stretch and never be motivated from within.

Separation

It is an achievement for the baby to feel he is separate, but the next step, which is separation, can be difficult for both parents and baby. There are many moments of separation throughout our lives; some say the first occurs at birth when the baby leaves the mother's body and the unbilical cord is cut. It is certainly the first instance of the body needing to use its own resources in order to start breathing. Until then the mother's body has also processed the baby's food and waste for him. Each beginning and ending – of a feed, sleep, gazing at each other – is a mini coming together and separation.

BEDTIME

Many babies are rocked to sleep, walked up and down, taken out for drives, all in a bid to get them to sleep. The parents' attitude and the baby's personality are both significant. If you have a rather sensitive infant, who startles at the slightest movement or noise and cries easily, you may be tempted to tip-toe around, especially if he is asleep. As we saw at the beginning of this chapter, one of the parents' roles is to translate the baby's emotions into something he can cope with; it is also to give him a picture of the world. A mother who is very careful not to disturb a baby, or anxious about waking him, will reinforce his sensitive nature. She will adapt her world to his wishes, but this will be impossible when he ventures further. I am not saying that sensitive babies should be bombarded with noise, but that we should think about what is a reasonable amount of chaos or liveliness in our household, considering other members of the family and what is manageable for the baby.

When a baby is drifting into sleep, he is leaving behind the world of those who are awake around him. He has to let go. Our own feelings about letting go will strongly influence our approach to this time. Most parents fear sleep a bit, especially in the very first days. Many a father or mother has gone to check whether the quiet baby is still breathing. There is a strong link between sleep and death. Pets are 'put to sleep' and people may talk of someone who has died as having 'gone to sleep'. (This is not a helpful way to explain death to children!) Death and sleep are tied in our unconscious, and in much of literature:

> Our very hopes belied our fears,
> Our fears, our hopes belied.
> We thought her dying when she slept,
> And sleeping when she died.
>
> THOMAS HOOD, '*The Death Bed*'

In sleep we may feel isolated, inaccessible. We may fear that we will not return to those whom we leave behind, or that they will not be there for us when we awake. Our capacity to part from our baby may be affected by difficult separations in our past, or more mundane factors to do with working all day, missing the baby, excitement at being reunited, loneliness and many more. All of these influences will play a part in how we present the forthcoming nap to our baby. A confident belief that sleep is restful, a lovely place to visit, a secure space within the ordinary day, will help a baby. A routine before and after sleep may also help make it an easier time which feels integrated rather than totally cut off from the rest of the baby's experience.

Ms T is a young mother with her first baby. Her partner is often away for extended periods of time on business and she has no family nearby. Ms T has many friends and is busy during the day. She is besotted with her baby daughter, Shona, aged nine months. They spend all day together, very much enjoying each other's company. The baby is cheerful, active, sociable and easy-going. However, nights are difficult. Ms T feels quite lonely. Shona will happily fall asleep at the breast but wakes as soon as she is put in her cot and cries. Ms T has never let her fall asleep by herself and Shona is used to falling asleep in her mother's arms. Ms T cannot bear to hear Shona cry, thinking that such a happy baby would not cry if she did not have a good reason. So they get into a pattern of Shona feeding, sleeping for a couple of hours, waking and feeding again throughout the night. This persists for months, even when it is clear that Shona no longer needs nutrition from the night feeds. But Ms T can see no way out, no solution which does not involve distress for her baby and therefore for herself. A close friend realized that both Shona and her mother were just exhausted. She moved in for a few days, to help Ms T while her partner was away. She stood by her as Ms T put Shona in her cot and let her cry herself to sleep. At first Shona cried for an hour and each day the time got shorter. After five days, Shona managed to drift off to sleep by herself in a few minutes.

The major elements which helped Ms T in this situation were two-fold. The first was the realization that both she and Shona were suffering from exhaustion. Although Shona was always chirpy early on, by about six o'clock she became increasingly fussy and irritable. Once she knew her reluctance to separate from Shona and to let her cry was not helpful, Ms T was determined to change. The second essential factor was the support of her friend. It had not been possible for Ms T and Shona's father to make the stand together. He was away too much, exhausted and preoccupied himself when home. It is not only single mothers who deal with such situations. Ms T's friend reassured her that Shona's cry was not desperate and sat with her as she too cried in the room next door. Much to Ms T's surprise after the first night, when Shona woke in the morning, she was her usual friendly self, and showed no signs of hating her mother.

We can understand Ms T's reluctance to separate from the baby she loves; she feels lonely without her partner or family. She also finds it hard to inflict what she sees as pain on her friendly and cheerful baby. However, from the baby's point of view, the difficulty in separating is compounded: the harder her mother finds it, the more impossible it seems to Shona. In the end quite a drastic limit was set. They both survived and, as they had a strong base together, it was relatively easy to recover. They also both gained in strength and did not need to huddle together quite so much.

The baby who falls asleep in company may be alarmed to wake up alone. He has not mastered the separation by drifting into sleep by himself. He may worry about falling asleep again. The baby in his cot will devise means to get himself to sleep and may very much enjoy this private time alone, as we saw with Michael's waking up alone.

Another opportunity for the baby to begin to form inner resources is during sleep. Sleep is not a stable state which we just go into and then awake from. Many emotions and sensations go on. One of the joys of Infant Observation is to see how much happens while the baby is asleep. Many an observer is amazed by his capacity to settle himself:

Tally, thirteen weeks old, is asleep. She is on her back with her head slightly turned to the right. She is absolutely still for about five minutes. Then a slight flinch around her mouth quickly passes and is followed by a wave of activity throughout her body. She draws up her knees, her arms and hands rub her face and ears. Sometimes she clutches the side of her head. She wriggles and twists her body. Her face puckers in distress and the movement culminates in a moment in which she looks as if she will awake and scream. She holds her head, one hand on each side, and I am reminded of Munch's *The Scream*. Her face unpuckers and changes into an expression of relaxation. Later she kicks and moves strongly and makes sounds of distress. She rubs her face and her thumb and mouth find each other. I hear sucking sounds. She moves both arms and the thumb falls away. This is repeated several times. Then with a more purposeful movement she places her right hand across her face and seems to get her thumb in her mouth quite quickly, where it stays. She sucks and settles back again.

We can see how well little Tally coped with what came her way and how she found means to stay asleep. Saying no to our urge to rush to the baby who whimpers creates a space for growth.

WEANING

Weaning represents another stage of separation.

Josh, twenty-two weeks old, has been breast fed. His mother, Mrs E, is due to return to work soon and in preparation is trying to wean him.

Josh starts to fidget and grumble as if he is building up to a cry. Mrs E talks about weaning the baby, who is now eating twice a day. She says she is going to try him with some home-made fruit juice today. As Josh is clearly unhappy, she goes to get the juice. She places him on her lap. He is now crying, screwing up his face and making lalala sounds and wriggling around. She talks to him but the crying is insistent and he is dribbling. She tells him the juice will help him and tries to give it to him. Josh brushes his mouth against the teat, takes a little and pushes the teat out with his tongue. His face is screwed up and red, his eyes are shut and he is making more noise than I have heard before. Mrs E is agitated and tells him not to cry. She tries to soothe him in her lap, then on her shoulder, and then to distract him. She offers him a number of things: the juice, her finger in his mouth to bite on, toys, a rattle, a ball . . . Nothing interests him. She offers him the dummy, which is also pushed out. She stands up with Josh on her shoulder. He is now screaming and she looks embarrassed. She says repeatedly, 'Please don't!' and talks quietly to him. She tries a teething powder which she says sometimes soothes him. Josh tastes it and screams yet louder. She tries the dummy again and then says with an embarrassed air, 'Last resort,' and offers the baby the breast while telling him there is nothing there. This immediately quietens the baby, who now has his eyes closed peacefully. In seconds he seems to be asleep, although he is sucking very weakly. Mrs E says, 'You're not using me as a dummy, you have a dummy for that.' After two minutes she prises him off the breast and quickly puts the dummy in. Josh starts to cry as the breast is removed but this time accepts the dummy and returns to sleep, very still.

In this example Mrs E knows she has to prepare for a return to work, she has enjoyed a very close intimacy with her son and has been talking to the observer about her worries about leaving her baby. The feed is important for both of them; it is not just the baby who is reluctant to wean. She feels pulled both ways – to let go and help Josh separate and to hold on and relish the last few days of breast feeding. The encounter is tinged with mother's feelings about having to speed the weaning up. At this point they may both feel unready for a premature separation. Guilt also lurks in the background and in my experience it is one of the great stumbling blocks to clear thinking. We also see a common scenario of the baby not accepting anything else, getting quite distressed and refusing a substitute. However, the comment about not being used as a dummy indicates that Mrs E is telling Josh quite firmly, in words at least, that he does not actually need milk from the breast, which is in fact empty, and should make do with the dummy for comfort. This introduces the idea of a substitute carer too; does he really need her all the time, or will another, the childminder, do for some of the day? The struggle in weaning is clear and Mrs E does not follow through her wish to wean and does give him the breast. Her own pain at leaving him and giving up feeding, added to his reluctance to let go, take over. At this point there may also be some pride in knowing that her baby wants her the most, 'breast is best'. By the end of the feed, she does make a successful attempt to help Josh accept a different form of comfort.

In always giving what a baby asks for, we are agreeing that the baby's choice is what is best for him. Here it would be equivalent to saying, 'You are quite right, nothing is good enough except the breast, I'm just fobbing you off with the juice.' Unless you can believe that what you are offering is good, you cannot present it positively. At this point juice is a new taste, the baby may be reluctant. If you agree that nothing else will do, you are not introducing your baby to the other pleasures of life – different tastes, smells, textures. You are saying only the familiar is ever good. The pioneering child analyst Melanie Klein stressed that it is important to remember that weaning is also weaning *to* and not only *from*. With 'weaning from' it seems as if there is everything to lose – the warmth, comfort, cosiness and closeness, the feeling that this is exclusive, nobody else can do it. There is much at stake for mother and baby. In 'weaning to' you have the whole world as your playground, so many new things to access.

There is also the question of rivalry: will the baby eat other people's food as well as mine, will he like the childminder or nanny better? There may be a tremendous sense of loss, but there is also a great opening. The mother has the opportunity of seeing her baby developing, growing into his own person and having the joy of a new relationship between separate people. For herself, she has the opportunity of more personal freedom, less neediness of her specifically, she can reclaim part of her body as well as her mental space.

Why is it Difficult to Say No?

The way we react to a baby's protests is often affected by the feelings that arise when we cannot have what we want, or when we are faced with another person's protest, anger, disappointment or persistence.

CRYING

After the birth of a baby, mothers, particularly, but fathers, too, can experience very primitive and raw feelings. They are having to re-create themselves, to play a new role. They may remember feelings they had towards their parents. They have been wondering what kind of parents they will be, and perhaps imagining the baby's character too. When the baby is dissatisfied or cries, most parents find this quite difficult to hear and bear. They may also feel criticized. The cry hurts them and they want to make it better. Often they do not know how to do this, so they look to the baby to let them know the solution somehow. Then we have two or three distressed and insecure people in the room instead of the one, the baby. It is at times like these that the parents need to distance themselves from their baby, to separate his feelings from theirs. They need to think for themselves rather than look to the baby for guidance. It is their translation of his cry that will help him put a shape to his upset and so start to find a way out of it.

Crying is probably the baby's most effective mode of communication. Research has shown that parents can recognize their new-born's cry from others by the third day. By the end of the second week or so they can differentiate between their baby's various cries. Parents are proud of recognizing which cry means what. At other times it is not so easy. A parent needs to absorb the emotion, feel it, before responding to it, in order to understand it. Often the cry goes right into you – it is meant to, this is how you come to understand what it means. Only then can you translate it into something more palatable for the baby. I am not saying this is easy; we even have physiological reactions to certain crying, which are the same as when we are faced with an emergency (an increase in adrenaline, higher blood pressure and more oxygen to the brain).

Our own experience colours our reaction, so let's look at emotions and thoughts which may cross our minds when our baby is crying.

- Does the cry sound reproachful? This may feel like a judgement on your care and you may respond by being hurt that the baby is saying you are no good. Does this echo internally with how your husband sometimes treats you? Your son is 'just like his father'. You will be tempted to give up trying to please and become withdrawn or angry.
- Maybe the cry reminds you of your stubborn and demanding little sister, who just had to screech to get her own way. Then you will react differently, probably crossly; the boundary between your baby and your memory of your sister becomes blurred. You could ask yourself: who are you really reacting to?
- The baby sounds full of rage. You feel guilty and wonder what you have done wrong. You become uncertain and hesitant and he feels more insecure.
- The baby seems in agony, you panic that he may be terribly ill. You feel all flustered and cannot think what to do. Your mind races ahead and already you imagine the baby might be dying. The infant himself may well feel he is collapsing when in pain. He needs you to tolerate the worry for long enough to think about it and check out what is really the matter. If you get swamped by his distress, you add to it.
- Does it sound as if the baby is falling apart and distraught? Does this fill you with despair? You may be reminded, perhaps not by clear memories but by feelings, of how helpless you felt as a child, unsure if things were ever going to get any better. The feelings of isolation in such a situation may flood you and then the cry is less your baby's than your own.

The following observation is an example of how a mother's feelings merge with her baby's cry. During a feed the mother had talked about a very long and painful delivery which left both her baby and herself with physical problems.

Jane (two weeks old) is restless; this turns into a cry, then some prolonged snuffles and wails. Her mother picks her up and moves her up and down with a slight bounce, talking to her all the time. The movements turn into a vigorous rocking. Jane grizzles and her face screws up. This feels very uncomfortable and dissonant to me and I feel anxious. Too much, too much, I think, and want to rock the baby gently. I wonder who is being comforted and rocked: the baby or the mother?

Here the cry seems very difficult for the mother to bear. In rocking the child, it feels as if she is trying to rock both their discomfort away; the physical movement is a way of trying to shake off the mother's feelings about the delivery.

How we manage our feelings about our baby's crying will have an impact on how he deals with his emotions too. If we give in to panic, we will reinforce his fears. If we ignore him, he may despair, as he cannot cope by himself. He may eventually give up protesting and become withdrawn. By trial and error, we will have to make sense of the baby's cry and try to offer what comfort we can. Sometimes this involves something we can do, and we usually go through the whole gamut of what he might need – a feed, a cuddle, a nappy change, a blanket, or whatever. At other times the baby needs to find his own solution. Having checked out all that we can, we need to say no to our attempts to make him better and to his crying call. The eminent American paediatrician T. Berry Brazelton writes that:

> At a time of day when the baby's nervous system is raw, parents' over-anxious ministrations can overload the baby's capacity to take in and utilize the stimuli. Colic can result . . . Constant handling may even interfere with the baby's own pattern of self-comforting and self-consoling . . . When he needs to be left for short periods to 'cry it out', they are likely to overdo their attempts to quiet him.

As we have seen all along, the baby needs some space to himself, a time to reach for what he needs, on his own. It can be very difficult for parents to allow this, to restrain themselves and not intrude on the baby's own efforts. They have to deal with their discomfort, not just the baby's.

UNTANGLING FEELINGS

Working out who feelings belong to or what they are about is complicated and the regular subject of many psychotherapy sessions. The following example comes from my work as a child psychotherapist in a hospital paediatric department. The consultant paediatrician referred little Zuleika, aged fourteen months, with her mother because she was grossly underweight for no medical reason they could find. She was also very pale and floppy and not getting better.

Mrs C was anxious about her baby not thriving and exhausted because Zuleika was breast feeding a lot, especially in the night. When we spoke, it became clear that Mrs C and Zuleika were very close physically. Mrs C sat stroking and holding Zuleika on her lap, noticing her every little whimper and not allowing her to stray too far. Mrs C described being away from her family, who lived abroad. She felt her marriage had changed since the arrival of the baby – her husband resented that she was less available to him. He worked long hours, was not very helpful in the home, being rather traditional in role allocation. He missed the intimacy and sexuality they had shared before and she felt he could not sympathize with her need for rest and support. As she talked, she got very tearful, Zuleika leant into her chest and Mrs C hugged her baby close. I pointed this out to her and we talked about her missing her own mother and her need to be 'fed' good things so as to be stronger for Zuleika.

In our sessions she was quickly able to link her feelings towards Zuleika with her own need for comfort, holding and feeding. We noticed together how Zuleika reacted by cuddling her or patting her when Mrs C talked of her sense of isolation. We also discussed how huddling up with Zuleika, holding her close, made Mrs C feel better, less lonely. The intimacy of the feeds also made her feel connected.

In this way, she began to differentiate her needs from Zuleika's and very soon Zuleika made progress. My understanding of Zuleika's experience is that although her mother fed her continuously, this did not feel emotionally nutritive but rather like a drain on Zuleika's resources. During the feeds Zuleika provided consolation and solace to her mother, but this left her feeling depleted, therefore pale and floppy.

Distinguishing whose needs are being met can be very helpful and often achieved in brief work. I saw Zuleika and her mother twice and know from the paediatrician that progress was sustained. I am a firm believer in early intervention to help parents at this very vulnerable but also exceptionally flexible time, when so much else is changing in their life. There is a chance of altering patterns before they become established.

We have seen that, especially in the early days, mothers are put in touch with their own infantile, often needy, feelings. They may want their own mothers nearby, to look after them and also to give their blessing to their daughter in her new role as a mother. Many cultures have rituals which ensure that new mothers are mothered themselves and not over-burdened by having to look after the entire family.

GHOSTS IN THE NURSERY

We all carry in our minds and hearts all sorts of people – parents, siblings, friends, teachers and others – with whom we have internal conversations. Sometimes they are helpful and sometimes not. Particularly when we are under stress, these figures come to the fore and we are aware of them. For instance, when struggling to soothe your fretful baby, you may gently stroke his brow, remembering how good it felt when your mother did that for you. You may consciously recall this and repeat it or you may automatically do it, without really knowing why.

Sometimes the impact of our past is far more unconscious and not so helpful. A great American child psychoanalyst, Selma Fraiberg, wrote:

> In every nursery there are ghosts. They are the uninvited visitors from the un-remembered past of the parents, the

uninvited guests at the christening. Under favourable circumstances, these unfriendly and unbidden spirits are banished from the nursery and return to their subterranean dwelling place. The baby makes his own imperative claim upon parental love and, in strict analogy with the fairy tales, the bonds of love protect the child and his parents from the intruders, the malevolent ghosts.

Mr H found it quite difficult to adapt to his new role as a father. He missed time alone with his wife and was jealous of the intimacy she shared with their new baby son. He felt very guilty about this but could not control it. He resented being asked to help with the baby and this caused friction in the marriage, which led to further irritation with his baby. A spiral of bad feeling was building up.

When they came to see me, Mr H expressed very clearly his jealousy of his son. He complained to his wife: 'You do everything for him and hardly notice me!' He was also cross about having to fetch and carry for him, saying, 'Why should I, I get nothing from him.' Mrs H would get defensive and try to justify her behaviour, explaining that at this stage the baby needed her more. I was struck that she was speaking to her husband as if he were a rivalrous sibling rather than the father of their son. He, too, spoke in a rather childish manner. When we discussed this, he was reminded of how jealous he had been of his little brother and what a handful he had been for his mother. Once this was out in the open, they were both more alert to signs of slipping into patterns of the past. Mr H tried to see when he lost his son from view and instead saw his brother. Mrs H resisted the temptation to treat her husband like a child. This was no miraculous cure, but the beginnings of an untangling of past from present.

Ms J found her four-month old daughter, Mary, very demanding and difficult. Mary would frequently cry for hours and not settle. When this happened Ms J felt inadequate and helpless. When we talked about it, she became tearful and spoke of how she tried her best and nothing ever seemed good enough for Mary. I commented that this seemed to distress her deeply. She nodded and said that she had felt that way all her life. She had tried really hard, but her mother never praised her and always expected more.

We were able to see that Mary's cries echoed her mother's childhood experience and then Mary became like Ms J's exacting mother in her mind. This led her to fuss over Mary excessively and to try to get positive responses from her. She yearned for a reflection in her baby's eyes that she was a good mother. This, however, put too much pressure, too many demands, on Mary, who became overwhelmed and cried. With time Ms J was able to see Mary's cry as needy rather than critical and was far better able to deal with it.

BEREAVEMENT

In some cases the 'ghosts in the nursery' relate to an actual death. One mother I saw found it very hard to separate from her baby, Ali, even for a short time. Mrs O was not comfortable letting anybody else look after him. She also had great difficulty in setting any limits. In our first session, she cried and recalled her experience of having lost her previous baby when he was three weeks old. She had wanted to name this son the same and saw him in some ways as a replacement. The wish to replace is very common, even after a miscarriage. When she looked at Ali, she did not always see him but saw instead little Ahmed, who had died. Her urge to protect Ahmed and not let him die became transferred to Ali, who was seen as fragile and at risk, even though in reality he was sturdy and healthy. This is a very understandable reaction and by tracing it back to its roots, we were able to look at Ali himself and think about what *his* specific needs were. Mrs O was also able to realize that by seeing Ahmed instead of Ali, she was not helping Ali. Her behaviour was having the opposite effect to what she was aiming for. Ali, instead of growing up with the secure reflection in his mother's eyes that he was a strong boy, saw himself as fragile and needy. What was done to protect Ali in fact made him more vulnerable. The work we did involved observing Ali closely together and thinking about what he himself was like, how he reacted. Establishing Ali as a separate individual with his own reality shifted the focus from Ahmed. The live child became more real and Mrs O was then, by contrast, also more in touch with the loss of Ahmed. Ali would always be Ali and never Ahmed. Having to let go of the dream of replacing Ahmed also helped Mrs O to mourn him. Once she was freer to see Ali as himself, she was able to be firmer with him, to say no when appropriate, without the fear that he would be overwhelmed or hurt. His sturdy nature was then allowed to blossom.

Another mother I saw following the death of a previous baby presented a different picture. I had worked with Ms C when her baby daughter, Angie, was in the Special Care Baby Unit. We carried on our meetings following Angie's death and through the next pregnancy, birth and first few months of her second baby, Cassie. Angie had survived, although very ill, for seven months. Ms C had been very attached and dedicated to her, visiting the hospital all day, every day. When she became pregnant again, she feared that she would not love the second baby as much. In the early weeks after the birth, she found herself feeling quite upset by Cassie's demands and unable to respond to them. It was easier for her to say no, than to give her what she wanted. She would talk about her as 'the baby', hardly ever using her name. She described her as a fat little lump, without much personality, sleeping and feeding, possibly a bit greedy. In contrast she remembered Angie as small but full of fight and determination. She had admired Angie's capacity to struggle. In talking about it, she felt angry that this baby, who was born healthy, should have what the other had not. She also felt disloyal to her first baby for giving to this one. Finally, she was very frightened of engaging and becoming attached to the new baby for fear of re-experiencing the tragedy which had occurred before. She was ridden with guilt about her reactions towards Cassie. By being able to voice these feelings, which she felt were totally unacceptable, she made them available for thought with someone else, rather than just being paralysed by anguish. We were then able to separate Angie from Cassie. They may not have the same qualities and Cassie was very different from Angie. We also discussed how making a space in her life for Cassie was not erasing Angie from her heart. The richness of care she had given Angie was still with her, it was hers to grant, and need not die with the baby. We worked very hard together to make a space to mourn Angie and eventually to welcome Cassie.

It is now well researched that after having lost a baby, grief and mourning are often delayed and certainly reawakened with the birth of the next child. Making peace with the ghosts of painful past experiences is a brave and arduous task. My heart and admiration went out to Ms C in every session and between for her courage, honesty and resilience.

Beyond Mother-and-Baby

FATHERS AND OTHERS

In the baby's first year, he is likely to be closest and most attached to his mother. However, fathers, too, play a crucial role. They provide a different way of being with their child, another intimate relationship. We have seen that mothers can get terribly entangled with their children and at times they may feel as confused and overwhelmed as their infants. The father plays a pivotal part at those moments, in helping the mother to retain her position and not get swamped by her infantile feelings. The father can protect her by coming between her and the baby she cannot separate from. This allows her time to recover, rest and have some space of her own.

In the early days the father can be seen as supporting the mother so she can support the baby. He can take the baby away from the mother when she finds it hard to say no, he can push for the baby to sleep in his own bed, he can offer a different opinion when a mother is trying to work out when to wean a baby, for instance. This is the role of father as gate-keeper. (Of course in some families the father is the one who finds it hard to set limits. What is important here is having access to an alternative point of view and a fresh outlook when we are tangled up with a baby.)

Research described by the paediatrician T. Berry Brazelton shows that fathers interact quite differently to mothers with their babies. Fathers tend to be more exciting, physical and playful, and babies prefer play-type activities with them. This is a positive attraction for the baby and a taste of relating to others.

As a baby begins to recognize his parents, he has a relationship with each one, much as if he formed a couple with one and then the other. In the triangular relationship of baby, mother and father, there will be times when he relates to both of them but he will also notice that they have a relationship with each other which excludes him. This is the beginning of him moving out of the couple with his mother and into the wider world. A baby is naturally egocentric and slowly learns that there are relationships which do not revolve around him, and eventually some which do not even involve him. There will be times when a father or mother says to the calling infant, 'Wait, I am talking to Mummy/Daddy.' It is an important lesson that what somebody else does is independent of you. Here the waiting is not because of what you have done but because of what they are doing.

I am writing of the father as the third person in the family because he is likely to have a major impact. Of course there is also the relationship with siblings, but they are usually monitored to some degree by a parent. I am also aware that many babies are growing up with single mothers. In that context, I believe there is still a great need for a third, an adult, so that the mother-baby couple do not stay enmeshed in a way that hinders development. That third may be a partner, a mother, a friend. It may have to be more than one person performing different functions.

CHILDCARE

One of the major milestones which occur at this stage is the transfer or sharing of care of the baby. All the topics covered in this chapter will have a bearing here. This change involves a separation, a weaning to another person, a reaction to the baby's protests and the issue of different styles and limits.

Choosing somebody to look after your baby can be agonizing. Do you select someone similar to yourself? If so, will differences seem more problematic? You may assume she will do everything just like you. Do you choose a young girl, whom you can instruct and teach? Will you worry whether she is responsible enough? Do you go for a granny figure whom you can trust? Will it be difficult to tell her what to do? All these choices will carry emotional baggage. You will relate to the person in charge of your baby with mixed feelings. It is just as well to be prepared for this. You may be filled with a *mélange* of gratitude, jealousy, relief, competitiveness, co-operation. It is important for the baby's welfare that you form a good alliance with his carer, and also remain close to him.

Sometimes when they return to work, mothers feel they have lost their baby. They see him as a little stranger at the end of the day, not having witnessed his moods and activities. If he is difficult, they may feel he has changed or that the carer is not good for him. It is important in these situations to ensure that communication is good. Brazelton carried out a study of four-month-olds placed in day care for up to eight hours a day. The setting was selected for its high standard. The babies tended to sleep and wake in regular cycles, and were never very intensely involved with their caregivers. At the end of the day, when the parents collected their baby, he seemed to collapse, wailing and fussing. The staff would inform the parents that he had not been like that at all during the day. Brazelton points out that the baby saves his passion, his intense feelings, for those who matter to him. It is a healthy sign, it means that the baby is attached to you when he complains after a separation. The important factor is for the two of you to recover.

It is unhelpful to jump to premature assumptions, about him or his care, about whether he no longer loves you, and so on. A child who cries upon meeting you can fill you with guilt, which interferes with your capacity to be clear. You may think, 'Oh, he has had a terrible time, I should never have left him, he will never forgive me.' You then turn yourself into somebody bad. This will be picked up by the baby who may then become cautious about closeness. If you have done all you can to ensure that his care is good enough when away from you, a more helpful approach would be to hold him and reassure him: 'Don't worry, Mummy's back now. I have missed you too. Yes, the day has been long hasn't it? But we are together now.'

You may be drawn into blaming his care-giver for all his grumpiness or changes in behaviour. You may have to stop yourself allocating blame, and think instead of what his behaviour means and how you can help. If you mistrust the person you are handing him over to, your baby will feel insecure and find the time away from you harder.

We also have to hold on to our wish to feel better and not push the baby to reassure us. You will try to soothe him, of course, but you also need to acknowledge his fury and allow him to express his complaints. If he is not able to express his upset and you do not help him process it, he will find rigid and defensive ways of coping. These will restrict his development.

We, too, have to grow in flexibility. At times, we will be tempted to make a distance between ourselves and our baby, in order not to miss him too much. We have to learn to bear the separation but also the being together – the coming and going.

We need to adapt to the idea that there may be many ways of caring for our child. Our model may not be the only one. We need to say no to our wish to control everything that happens to him while we are away. Of course it is important for you, as much as for your child, to clarify your rules, your boundaries, with the person looking after him. Beyond that you have to hand over his care in the belief that the other is also good enough. As we have seen, for people to get on with babies, the exchange has to be honest, to come from inside. Babies, particularly as they do not have language, pick up emotional communication, the look in your eyes, the tone of your voice. The way the care-giver relates to your child, whether she is sensitive to his cues and able to hear and manage his protest, will be a better indicator as to how well she will get on with him than whether her limits are exactly the same as yours. She will care more genuinely for him in her own style than if she is forced to adopt yours.

Sometimes, parents have little choice over the care of their child, but if he remains distressed, stops eating or sleeping, becomes withdrawn over an extended period of time, you will know that these are not ordinary reactions to separation. You may need to think again and find an alternative.

On a more cheerful note, your baby may settle very well, thrive and develop just fine. You will be relieved and happy but may need to curb your wish to be everything to him, the only one who matters. This is easier said than done. It may help if you remember that the baby has developed social skills and a capacity to be without you because of the care you have given and continue to give.

Summary

A new-born is a sophisticated being, very much involved with you, right from the start. In order to thrive, he needs you to be attuned to his communications. What matters most is the interaction and mutual responsiveness between the two of you. An infant can become easily overwhelmed by his feelings, physical as well as emotional. He needs someone who thinks about his state of mind. He needs you to understand what is happening to him and, through your reaction, dilute the intensity so that he returns to a state that he can cope with. It is through your interpretation of his behaviour that he starts to get a picture of who he is. It is also through your handling that he learns about how people manage strong feelings. If you are attentive and responsive to his cues, your baby feels held and contained. He will be more able to take things in and participate in the world around him. Your responses and readings of his state need to be 'good enough'; a certain degree of mismatching and recovery is helpful to development. It is from this basic position that he will be able to cope with limits being set.

Saying no, in all its various guises, is essentially about making a distance between a wish or a yearning and its fulfilment. Certain aspects of child rearing – separation, weaning, how to cope with crying – bring the issue of limits to the fore. Most of us do not find it easy to hear or say no. Many factors impinge on us, which may be related to our history, our present situation and how we see ourselves. Our reluctance to set limits may hinder the development of our child's capacities. I have tried to show how saying no is useful. It represents a gap or a space in which other events can happen. From this point of view, it is not so much a restriction as an opportunity for creativity.

2

Two to Five

When I bring you coloured toys, my child, I understand why there is such a play of colours on clouds, on water, and why flowers are painted in tints – when I give coloured toys to you, my child.

When I sing to make you dance, I truly know why there is music in leaves, and why waves send their chorus of voices to the heart of the listening earth – when I sing to make you dance.

When I bring sweet things to your greedy hands, I know why there is honey in the cup of the flower, and why fruits are secretly filled with sweet juice – when I bring sweet things to your greedy hands.

RABINDRANATH TAGORE, 'When and Why'

The Magic Kingdom

A young child who believes there is a lion under his bed or that the toilet will swallow him up is not mad. He is an ordinary toddler. In his world, the difference between fantasy and reality is blurred. The child will wonder: do dragons really exist, are there such things as witches, are thoughts powerful, can you make things happen by thinking them? Young children are passionate and see and experience the world in extremes, very black and white, the shades are not muted or mixed.

All cultures have fables and folk tales which match this view of the world. Usually the good and evil characters are exaggerated and typically one-dimensional. The hero is all good, the villain all bad. Families have a beautiful daughter and an ugly one. The mother is perfect, the stepmother wicked. One brother is intelligent but unkind, the other stupid but gentle, and so on. Children love to hear these tales, over and over again, anticipating frightening or exciting moments, joining in, getting scared, being reassured, identifying with the hero and triumphantly cheering at the happy ending. We all connect with these pleasures, whatever our age, as the stories deal with essential human longings, hopes and anxieties. The search for true love which conquers all, as in *Beauty and the Beast*, the victory of kindness over tyranny in *Cinderella* – these morals encourage us to strive for positive goals, and reassure us that siding with our better nature will be rewarded.

Bruno Bettelheim wrote a fascinating book called *The Uses of Enchantment* which looks at the importance of fairy tales. He believes they help children to work through their fears and concerns. I am sure we have all witnessed a child, or remember ourselves, being fascinated by the scary parts of fairy tales and needing to go over them again and again. The moral of the tale is always reassuringly clear, although my father relates the joke of the little girl telling her mummy how frightening the story of *Snow White* is. The mother asks her which part was particularly alarming and the little girl replies, 'At the end, when that stranger comes and takes her away from the dwarves' house.' We choose to hear the story as we wish it to be. We do not listen to it with reality in mind, but with our imagination, and focus on how it touches upon our wishes and dreads. We would not find it alarming that the prince should instantly fall in love with someone he has never met or that Snow White, on waking, should give up all that is familiar to leave with this stranger. That is not the model we think we are giving our children in recounting the tale. We are telling the story that one person's love can save you from the attacks of others and rescue you no matter how dire the circumstances.

And so children grow up believing that parents are like the magicians and fairies who can make everything fine. However, they are still frightened, no matter how loving the home. The monsters, ogres and witches still lurk. Many families who try to keep everything gentle do not buy violent toys for their children, and then are shocked to see them enact great battles or be petrified of their soft fluffy bunny.

One aspect of children living in a world of magic is that they feel anything is possible. They are very literal and take words concretely and as they sound. Little Harry would fuss terribly when walking past a particular room in his house. Finally, he confessed to his mother that he was petrified of the terrible giraffe she had said was coming through the door. She was most puzzled until she worked out that she had complained about the draft! This is why it is important to be clear and simple when talking to young children. Threats like, 'Don't make that silly face. If the wind changes, you'll be stuck with it,' are quite real to small children. One of the reasons they may believe fantastic things is that their inner world, the world of their thoughts and feelings, can be very passionate. When they hear someone say, 'I was so angry with him, I could have killed him,' this is quite credible, because the child may often feel murderous towards those who thwart him.

It is easy to see why at this age the fantasy world and reality can be so intertwined. The fear of parents who are angry, the ease with which they are turned into monsters, can make a small conflict into a major event.

Jack reacts to his father saying no to him as if he were a giant out to kill him, just like in the story of *Jack and the Beanstalk.* Mr B is a gentle man who finds it hard to say no to Jack. When he does, Jack reacts with such passion, Mr B feels his son sees him as wicked. He tries to reason with Jack, who only gets more inflamed and acts as if Mr B would hit him, which he has never done. Mr B then gets quite angry inside himself with Jack for the image of himself he sees reflected in Jack's eyes, that of a nasty, brutish father. What starts off as setting a simple limit turns into a big argument and upset for both of them.

Jack is reacting to his own interpretation of what a 'no' means, rather than to what his father is actually like. His world is of fairies and witches, wizards and ogres. When Dad says no, he is being transformed by Jack into someone powerful who is against him. Jack himself probably wishes he could be a giant and stand up to his dad. By his behaviour, though, Jack turns his father, in his mind, into something he is not in reality. His father is then understandably upset and paradoxically may become much nastier. He may think, 'How dare Jack treat me as if I am bad or would hit him?' At that point it is likely that, rather than Dad reassuring

Jack, anger will prevail and Dad does then become like the giant, voice booming as he tells his little son off. So Jack's fantasy finds an echo in reality, and the two become blurred.

Setting Limits

It is the parents' challenge to nurture their child's passion and involvement in the world, as well as teaching him to fit in with society's rules. The issue of saying no becomes particularly prominent during the toddler stage. The child is mobile, into everything and many hazards are in his way, so the question of discipline comes into the open. In the ordinary life of a household with a toddler, 'no' could well be the most frequently used word!

It is difficult to be on the receiving end of a no, and we deal with it in a variety of ways. One child may use the word himself, as if wielding a weapon, shouting 'no' powerfully when asked to do anything. Another may use the word to identify with the adults. When my daughter Sushila was two, she found it quite difficult to be told 'no' to all sorts of things. She was on the whole very well behaved and did not really protest. However, when in the company of her fourteen-year-old cousin, she would respond to whatever he said by speaking in a rather patronizing, adult tone: 'No, Teshi, no,' as if he were really an unreasonable boy. I imagine that her way of coping with being faced by many noes was to take on the adult role as she experienced it, and make him feel as she did: silly and little.

Toddlers are usually impulsive, active, demanding and curious. All these attributes can be seen as qualities or faults depending on your point of view. A toddler pulling out pans on to the kitchen floor and banging on them with spoons can be seen as a budding drummer with a fantastic sense of rhythm, a scientist exploring the characteristics of objects coming together, a noisy little brat, a messy boy with no sense of what objects are for, and so on. Your picture of him will depend on many factors, too. As we saw in Chapter 1, so much goes to forming who we are and how we see the world. And then there is what has happened on that particular occasion. Our reaction at the end of an exhausting day will be very different to our response in the morning, after a good night's sleep. Another very important determinant is how we feel about the life we live and how the child fits into it. A mother with many forms of support will be more likely to see the humour of an event than one who is stretched to the limits of her capacity to cope. Whatever our reasons for behaving as we do, our reaction to the young child is a communication.

A mother and her two-year-old are at the supermarket. Johnny smiles at people around him and chatters pleasantly to his mother. Soon he becomes restless and his mother gives him a sweet to keep him happy. As she carries on shopping, he demands more sweets and she gets irritated, stating that this is bad for him. He starts to whinge and demand more, she gets cross and says he's had enough for today. He cries, other shoppers stare, his mother feels angry with them and with her son for embarrassing her. She gives in. He now wants other things, is squirming in the trolley and crying to go home. His mother offers more sweets but he is crying so much he does not want the sweets any more and throws them to the floor. His mother is furious and shouts at him.

Here, as with the babies, we can see a mismatch. Although at first little Johnny is happy to shop, soon his tolerance is stretched. The rigmarole with the sweets becomes a battle rather than a treat; the sweets are therefore no longer comforting, they have lost their original meaning of making him feel better. They become a bribe or something he has extorted. The exchange is fraught and exacerbated by other people's disapproval. His mother finds it hard to stick to the limits she has set and resents this.

In all such interactions there is more than one story. This may sound obvious, but frequently, in trying to help a family, only one version is heard, usually the mother's. First of all, it is important to listen. This is true not just for those of us who help families but especially at the time, for those involved in the situation. If a child is reacting with difficult behaviour – whinging or crying – this is a communication; in this instance it probably means that the shopping is lasting too long. Realizing this, Johnny's mother has various options: she can carry on regardless; she can stop; try to make the time more pleasant; get irritated that her plans may have to change; organize the next time taking into account her son's capacity to cope with this activity . . .

If we investigate, we often find that there are other reasons for how she reacts. Mum is stressed, she may know Johnny hates shopping but has no choice about taking him; she may feel unsupported; she may feel cruel for repeatedly putting him through something she knows he hates; she may think he's spoilt and should just put up with it; she may feel guilty about giving sweets which she knows are bad for him . . . This makes it harder for her to stick to the boundaries she has decided on. She is unable to tell Johnny the shopping has to be done and he will have to put up with it; she is also unable to find a way of getting him to join in, perhaps helping her to pick out items. It seems his crying paralyses her thinking and she gets into a battle with herself and with him.

To be firm, you have to believe that what you are doing is right or you carry little conviction and the child gets a mixed message. He may then believe that if he fusses enough, you will give in. We so often see this picture of parent and child whinging on at each other, caught up in a tangle of misery in each other's company.

THE PROBLEM OF CONSISTENCY

On the whole, adults have a greater capacity and understanding than children, and children look to their parents, particularly, to make sense of the world around them. In Chapter 1 we saw how this begins in infancy, with babies needing their parents to put a shape and a meaning to their feelings. As the child gets older the parents give him a picture not only of who he and they are, but also of the world at large. The healthy child will refer to his parent when meeting new people to see if they are safe to interact with; he will also use a parent as a base to explore and to get feedback about his activity.

So we will see and hear parents introducing friends, asking Johnny to smile at Aunt Mim, to go and fetch the toy, look at this wonderful book . . . He will be praised when he achieves something his parents are proud of, scolded when he does something that worries them. Through all this talk and play, Johnny gets a sense of the world and what he can and cannot do in it. If the responses to his behaviour are consistent, he gets a more solid and clear view, he gets a good idea of what is allowed and what is banned, what is safe or dangerous, what is frightening or not. No parent or adult is completely consistent, but a general picture does emerge for the child.

It is common knowledge that consistency is desirable, especially in setting limits, so let us look at what interferes with it.

The simplest reason for two parents being inconsistent is that they have differing points of view.

Little Laura, aged three, is a charming, engaging child. She spends the day with her mother, who, on the whole, enjoys being with her. She is active, friendly and welcomes other children's company. At night time, she is reluctant to go to bed without her mother or father lying next to her and staying till she falls asleep. Mrs M feels she needs her own time in the evenings and resents having to stay with Laura, whom she has been with all day. She is firm and tells Laura she will have to get to sleep on her own. Mr M doesn't really mind, it gives him a chance to be with Laura and to unwind from his day at work. It is not a demanding time, he does not have to play with or entertain Laura. It suits the two of them and he stays with her.

It does cause friction between Mr and Mrs M, though. They are missing out on time together as a couple, they argue about whether Laura needs this routine to help her sleep. Mrs M feels irritated that Mr M is setting up a pattern which she is then stuck with if he is away. Their different approaches cause difficulties between them. When they discuss why they each believe their point of view is right, they are faced with not understanding the other, and get irritated with each other. At that point if Laura cries for her father, he is far more likely to want to be with her than stay and argue with his wife.

We can see that what then happens between them and how they manage their differences has an effect on Laura. They cannot be consistent for her because they cannot agree. She gets a mixed picture about what it means to go to sleep. Initially, it is regularity and consistency which will give her an idea of what bedtimes are like. As she gets older she may learn that bedtime is different according to who is at home. In this example, there is tension around the whole issue and bedtimes become more insecure, thus exacerbating her reluctance to go to sleep by herself.

Mohamed, aged two, spends weekdays at his grandparents' home while his parents work. They pick him up every evening. On the whole the arrangement works well, but there are disagreements over certain details. His mother is trying hard to wean him off his dummy, which he has got used to sucking whenever he wishes. His mother and father agree he should not have it so freely available and want to restrict its use to bedtimes. They worry about it deforming his teeth, questions of hygiene, as he drops and picks it up all the time, as well as it becoming a habit. The grandparents, on the other hand, see no harm in it, they feel he misses his mother and needs the comfort. It also allows them more peace if he is contented. They cannot agree and Mohamed has a different limit set depending on who he is with.

Both sides have legitimate arguments based on their concern for the child. It may well be that Mohamed does not need the dummy so much or so often now, and his parents may want him to develop other means of coping. They are thinking of the part of Mohamed which is growing up. The grandparents' point of view puts the parents in touch with the needy and baby side of Mohamed. This will exacerbate their guilt at leaving him, reminding them that he misses them. There may be times when Mohamed can do very well without the dummy and others when he needs it. The polarization of his carers makes it hard to observe Mohamed's reactions and to act according to his need rather than following their own views about what he should need. He is caught in the crossfire of their determination to stick to their belief and their view of him. Mohamed gets a mixed picture when he asks for the dummy, and this confuses him too. Is he a little boy who is growing up, who hardly needs the dummy? Is he a small child who needs the dummy in order to hold himself together or comfort him because he is feeling bad? What is lacking is a thinking space for him to be a bit of both. He is not being helped in his struggle to separate from his parents, and from his dummy, in a way that is sensitive to his pace.

Mixed messages like these can have varied effects. An older and secure child learns over time that Mum and Dad can be very different from Granny and Grandpa. He will accept different rules in different homes. However, if he is not too secure or if there is a real struggle between the needy baby and the more grown-up parts of himself, he may feel more confused and insecure, never knowing if his request will be granted or not. In his mind, it would be as if he were always being told 'maybe'. He might feel very tantalized and become cranky and anxious, anticipating the withdrawal of the dummy even when it is given to him, and so asking for it and clinging to it more than he would if the adult rules were clearer. In Mohamed's case, he is still very much in between, crying for the dummy and getting upset with his parents when they refuse him, but slowly getting accustomed to coping without.

Such conflicts are likely to occur between two parents, between parents and those caring for their children, such as relatives, nannies, child-minders or au-pairs. However, different feelings exist not just between people who disagree or have dissimilar approaches and beliefs, but also within ourselves. So, to return to the supermarket example, we may feel mixed about whether or not to allow sweets; at times we may feel there is nothing wrong with the occasional treat but at others we think it is bad for the health. Other examples might include playing with food, being cheeky, going to bed at a particular time.

Let us imagine a situation similar to Laura's:

Tom, aged three, also wanted someone to stay with him while he fell asleep. His mother thought this was a bad habit and tried hard to be consistent and firm about it. However, at times, she remembered how she used to feel as a little girl, how big and cold her bed seemed and how lonely she felt at night. She would then have an internal argument about what she should do: she should be firm or he would never learn to sleep by himself; what harm would a little cuddle do?; he was becoming spoiled; why should children have to grow up so soon?; she had felt lonely but was fine now, wasn't she?; why shouldn't she spare him the pains she had gone through? On different occasions the argument was won by different voices.

Tom never knew if his protests would get his mother to stay or not. So he always asked, and fussed, and inevitably felt disappointed if she did not stay. The inconsistency makes for a more fraught moment, not knowing if your hopes will be dashed or fulfilled. The victory will also be less sweet because the prospect of the next such moment looms. Children are more comfortable with predictable outcomes, even if they are not what they wished for, than with the roller coaster of hope and disappointment.

In this kind of scenario, the first helpful step is to be in touch with what the child's protest triggers in us – in Tom's case, his mother's memories of disliking being alone. Once she recognizes this as her difficulty rather than his, she probably does not want him to grow up with the same reluctance to be alone as she had. She can then see Tom more clearly and find the means to deal more consistently with his requests. This could take the form of a plan which she explains to him – she might agree to spend half an hour every night with him, in his bed, reading him a story, before leaving him on his own to fall asleep. This would satisfy the wish for closeness and comfort for both of them, as well as giving Tom the idea of a limited time after which he has to devise his own way of getting to sleep. Tom's mother can reinforce this by offering a soft toy, blanket or a piece of cloth which belongs to her and suggesting he can cuddle this instead once she has gone. She is helping him to make a transition from herself to other comforters. It will be easier for her to be consistent in her limit-setting once she is secure that she is offering Tom tools with which to cope.

Whether we react consistently to demands, what stand we take or how we deal with conflict is highly affected by our own make-up and our history.

Mrs F had grown up in a strict family with a rather tyrannical father who always pulled her up about her posture, her manners and everything she did. She could see this characteristic in herself, in that she was always extra sensitive to her four-year-old daughter's faults and would often repeat with her the experience she had had – she would find herself behaving as her father did towards her. However, she was aware of this and felt guilty about it. Still, when she got into conflict with her daughter, the memory of her past arguments and feelings about her father were revived. It was as if she was reliving a disagreement with him rather than her daughter. Her own childish feelings became more prominent than her adult ones. This frequently filtered through to her behaviour, which also became childlike. It became difficult for her to see her daughter's actual reaction instead of remembering her own. They then got into battles of will like two toddlers rather than as an adult and a child. The relationship became distorted and old feelings were more active than attention to the present.

Mr R had a very deprived childhood, both materially and emotionally, although now he is wealthy and happily settled. Saying no to his children and seeing their objections stirs up his own overwhelming feeling of neediness as a child; he therefore showers them with gifts and treats, which they take for granted and do not really appreciate. Instead of having the desired effect of making them happy and contented, as he had not been, they seem dissatisfied and demanding. This makes him angry, because he feels they have so much more than him, and he then sees them as greedy and spoilt.

Again an unhelpful cycle is established. This is somewhat akin to someone who has been starved overfeeding his child: you can understand the reason for his reaction, but this is not relevant to the child's experience at the time.

These examples reflect how the past impinges on the present. As we saw in Chapter 1, it is as if figures or experiences from one's history come between you and your child. Sometimes these may be real figures you remember: your parents, siblings, other family members, teachers. They may also be your own self at different ages – you remember your experience of being the age your child now is, for instance. This means you are not reacting to the present with your actual child, but reliving something of your own. We have internal dialogues, arguments, with these figures from our inner world. If we are distressed we may call upon a helpful voice which reassures us and tells us that everything will be all right. When we make a mistake, we may feel chastized by a strict aspect of ourselves. This inner world is part of what makes us multi-dimensional and enriches us. It also causes the discomfort of ambivalence, of feeling confused, and of often acting in ways we would really rather not.

SANCTIONS

Sanctions are important when you are trying to enforce your no, but I do not think there is a recipe that works for everyone, so I have not made sanctions a focus of this book. If you have conviction about your stand, if you can be in tune with your child, generally an under-five will respect your word. Of course you will need to back this up at times. There are many strategies you can use, whether it is to reduce TV watching, send your child to his room for some time out, confiscate a favourite game, hold him physically when he is having a 'wobbly', refuse to take him out to the park if he misbehaves, or whatever. You are best placed to know what will have meaning for your family. It is not the sanction itself that will matter but what is communicated through your behaviour. You do not need a sledge-hammer to crack a nut. Heavy-handedness usually backfires, as does losing your own temper, humiliating the child and getting into a battle of wills. I do not think it is ever helpful to lose your temper; behaviour which is out of control is frightening for parent and child. However if at times, as happens with all parents, you do or say something which you regret, it is not the end of the world. It can let the child know you too are human, not a robot or an angel. This may let him see himself and his passionate feelings in a kinder light. If you do go over the top, an apology can also be very positive. You are giving the child the model that you can consider what you have done, realize it may have been wrong, admit to it and ask for forgiveness. That opens up these possibilities for him, too.

What is of value is to hold on to your role as an adult: to feel for your child and the state he is in, as well as being able to think about what is best for the two of you. You need to retain your own self-respect and convey to him that your 'no' has a reason. You do not have to explain every reason to him; it is sufficient that you know what you are doing. In this context I believe that the very occasional, small smack to stop an escalation of feeling and conflict may be preferable to a long tirade about the child's poor behaviour. Many a parent of this generation can burden the child with lectures and defensive explanations.

The point about sanctions is that they should help the child to learn. Cruelty only teaches a child to be nasty. Your sanction should be aimed at helping him be more thoughtful. You are likely to find out what works best for you by trial and error. As long as you have regard for yourself and your child, just the fact of trying to make things better helps. Children are deeply appreciative of people who struggle on their behalf. They know that it is often easier to give in than to strive for a better solution.

Never Saying No

One of the striking characteristics of toddlers is their gusto in approaching new tasks. They will often clamour to 'do it myself', much to parents' pride but also exasperation. So you will have

Alexandra insisting on doing up her own buttons, Jagdish wanting to tie his laces, Kai climbing on chairs to reach a toy, and Panayota pushing the baby's pram which is clearly too heavy for her. It is a time when parents may need to curb their urge always to do things for their child. Children need the practice, to master their growing physical skills, fine motor control, manual dexterity. They can be determined and resolute in their attempts to do things for themselves. It is crucial at this stage not to restrain their enthusiasm nor to crush their aspirations. This can be seen in a typical mother's comment to her little son, Bob, who is helping her carry in the shopping: 'What a little man, thank you, I could not have brought it in without you.' This makes him feel valued, strong like Daddy, someone Mother needs.

However, it is also important for children to have a realistic view of what they can and cannot do. So it might be good for Bob to try to carry a heavier bag which he cannot manage; he will still be valued for what he can do, as well as in touch with a limit. This would also spare him the sense of responsibility of *having* to be mother's little helper. In a family with a single mother, particularly with a son, fostering an unrealistic sense of power may be a burden on the child.

The child's own character and how you help him deal with the frustrations of not being able to do certain things will lead to how well he struggles with not getting things right straight away. Of course, he needs to try in the first place, in order to find out where his limitations lie, where he might need assistance. The child who believes he can do everything by and for himself will not be able to accept help, whether this be for physical things or with learning. Denying any dependence leads to being rather bossy and at worst a bully. Bullies are usually scared that someone might be stronger than them and that they might be on the receiving end of what they mete out. In a world of magic this is even more frightening.

The dramatic story of Paul, whom I worked with in a Young Family Centre for deprived families run by a Social Services Department, illustrates what can happen when dependence is avoided and limits are fought. I believe general lessons can be drawn from such extreme circumstances.

Paul was an unwanted child, the youngest of three boys. The atmosphere at home was explosive, passionate, violent and chaotic. All the boys regularly spent time in care. By the age of two, Paul seemed a hopeless case. He gave people the impression he was from another planet, eyes wild, never looking at you straight in the eyes. He was very over-active and agile, jumping off great heights in a dangerous way. He was given to sudden violent outbursts. He totally lacked concentration, was hardly verbal and very difficult to stay with.

In working with Paul, it was apparent that just being caring, kind and thoughtful with him was not enough. He exasperated even the most patient of staff members. He appeared to believe he could do anything and seemed not to care one bit if people were angry or punished him. I was in charge of his care at the centre but also offered him three individual sessions a week. This is a typical extract from one of our early sessions, after he had been away sick. He was four years old.

He hurled the phone into the bin and shouted, 'Fuck off!' He ran to the poster and pulled it off the wall. I said he was very angry with the room and me since he had missed so many sessions. He ripped and tore the poster before I could get to him. He flung a lamp down from the shelf. It did not break. I told him he might hurt himself and explained how dangerous it was to play with glass. He ran over to me, kicked me and shouted, 'Shut up!'

Paul gave the impression of being untouched by others' opinions or feelings. He acted as if he were omnipotent. The way he jumped, screamed and physically assaulted me made it very difficult to function as an attentive mother would – an experience he had never had. Sometimes I needed to hold him physically to stop him damaging himself or me; at other times talking about his feelings allowed him to control himself. It was a constant challenge to try to give his feelings and actions a shape through my recognition of them. Paul's entire behaviour seemed geared to denying any need, for anything or anybody. One of his favourite sentences was, 'Me do it.' In Paul's vision of the world, a mother could not be seen as a person who holds, feeds and helps him to grow. A father was an idealized brutal figure who hated babies and whom Paul identified with. This view spread to all adults. He had never felt safe enough to be a baby. It was a big risk for him to feel vulnerable because he might get hit or mocked. Much better for him to imagine himself as a rough and tough superman.

It was a very long time before Paul could allow himself to feel appropriately vulnerable for his age and before he could conceive that somebody might be interested in looking after him. He used hardness and callousness as a protective shell to stop him feeling open to attack. However this defence, like a carapace, also stopped good things getting in. Underneath, he was like a very fragile, frightened and unprotected baby. The only alternatives that seemed open to him were to be a harsh bully or a totally vulnerable infant. Being allowed to act so out of control and feeling adults could not stop his whirlwind behaviour supported the view that he was invincible. The baby aspects of him were then out of reach, not allowed to surface and be helped to develop.

With a child like Paul, it was important that an adult should insist that he was not invincible. Without puncturing his sense of himself, he needed to be reminded he was little, and he needed to feel someone's willingness to look after him. He also needed an adult to give him the confidence that he would grow strong, a hope that would counterbalance his feelings of fragility. As the sessions continued and boundaries became clearer, Paul started to feel more secure, to explore what it meant to be a baby who was held and thought about, remembered and cared for. It seemed possible for the first time in his life to allow himself to be vulnerable. We had a very moving session in which he picked up a doll and pointed to her eyes, nose, mouth and so on, saying, 'Look,' and, 'What that?' He finally put the doll down and said 'Boy' confidently. I felt he had been asking me about himself, what all the features were, what was a child, who was Paul. Just as a very young baby learns parts of himself with his mother (here's my nose, there's your nose), Paul had finally had the experience of being able to think about who he was and had come to the conclusion that it was OK to be a boy.

A more ordinary example of behaviour which leads to feelings of omnipotence can be seen with Charlie, aged three:

Charlie is a very demanding child who screams if he doesn't get his way. His parents are at the end of their tether because they feel they have no control over his behaviour. They tell me that he 'insists' Mother cooks pasta every day. He 'has to have' a story read before bedtime. He 'refuses' to let anyone but Mum put him to bed. The list of his demands is endless and the parents fit in with each and every one of them.

What is striking is that Charlie's parents feel they have no say in how he behaves, no recourse. They take his wishes for needs and comply. Charlie is a little despot in their home. This makes for unhappy relationships all around. He has no practice at dealing with frustration, and when confronted with a difficulty, he finds it increasingly hard to get over it. He has a bossy kind of identity. This allows no space for his more infantile side to be expressed and to grow. As his parents never say no, he does not have a taste of feeling furious, or that he will collapse, and then the experience of recovering. His development is stunted. He is also faced, like Paul, with the fear of retribution and the anxiety that if adults cannot stand up to him, he is left unprotected against someone stronger than he is.

At the other end of the scale you may have the child who is too good. The child who cannot bear to be little and copes by being over-identified in an imitative way with adults also avoids the necessary developmental pains of being little.

Anjeli, five years old, is a traditionally 'good' little girl: she is obedient, polite and quiet. She is quite forward in her learning and does well at extra activities, such as ballet. She has no tantrums and hardly misbehaves. However, there is a flatness to her. When she is at school, she follows instructions very well but is rarely seen to initiate a game of her own. It is as if she has put on an adult guise and is dominated by a wish to please those around her. She is seen as a bit of a goody-two-shoes by her peers and is not too popular.

As Paul adopted a Superman persona, Anjeli adopts a pseudo-adult identity. It is hard to recognize the child in her. She is spared parents' and other adults' crossness but misses out on some of the enjoyment of being a child. She by-passes the experience of being told 'no' by anticipating it and controlling herself rather rigidly.

There may be many reasons for Anjeli's behaviour. It could be that her mother has just had a baby and Anjeli feels pressured to grow up quickly, to spare her mother trouble. It could be that she picks up conflict between her parents and wishes to pacify the atmosphere. It could be that she enjoys the rewards of being her parents' good little girl and dares not risk creating a fuss. It could also be that she is more comfortable being quiet and out of touch with strong emotions. There are myriad reasons why we adopt the coping mechanisms and strategies we do. Each person's way has to be observed and thought about individually. For healthy development, though, we cannot by-pass some of the struggles of uncomfortable feelings. Martha Harris, an influential child psychotherapist, writes:

> The child cannot learn to control his undesirable, aggressive emotions unless he has had a chance to experience them, to know them at first-hand. This is the only way he can gauge their strength, the only way he can find resources within himself to harness them and, if possible, to utilize them to good purpose.

Feeling bad about saying no and being firm can lead to great problems, which I would like to illustrate with a rather extreme example of a little boy and his mother.

Darren was a three-year-old referred to a hospital paediatric department for chronic constipation. He was on heroic doses of laxatives which no longer worked. He had to be admitted to have the faeces removed under general anaesthetic and was likely to need further such interventions. In desperation, the family was referred to me. When I met him with his mother I was struck by how domineering he was and how cowed and shrinking she was. On talking with them, it quickly became apparent that Darren absolutely ruled her life. He would not stay with anyone but her, not even her husband or older children. He screamed when he did not get his way, and she always gave in to his demands. In their sessions with me he made it very clear that he hated sharing his mother's attention with me and shouted over our voices so that we could not have a conversation. Very soon he would want to go home and would scream at the top of his voice, pleading and begging to leave. His impact was so strong that, on one occasion, three different people on my corridor knocked to see if everything was OK. It was as if we were truly torturing this little boy.

My main task with his mother was to sit with her through this ordeal and to think about what was actually happening. We started untangling what was occurring in the room and comparing it with what she felt and what Darren was expressing. Was he really being tortured? Was it really unbearable to sit in the room and think about the problem? Was his mother really cruel to want to talk to somebody other than him? Was there no alternative? With time, and my insistence that we see the session through and not end early, he was able to calm down, to look at the toys and play a bit, and even to express how he felt about it all. He eventually played with the Plasticine and we were able to talk about how stuck he felt with the poohs, but also how he used them to get everyone else stuck, and how scary it then was for him when no one, not even Mum, could help him.

This was the more traditional aspect of our meetings – the interpretation of play which takes place in child psychotherapy. However, for this little boy I think the most helpful intervention was that I sat through his rage and fury with his mother and helped her keep to the boundary of our time together and so demonstrate to both of them that it could be done, they would both survive, and that it was helpful. She began to be firmer with him because she could see that instead of it being cruel, it in fact helped him. Reassuringly, after seven meetings, over a period of six months, the constipation cleared up, and he was freed from this stuck place where nothing could shift and thankfully did not need further intrusive medical interventions.

For children like Darren who totally dominate their mother or their family, life is not much fun. Even though they may not be left alone, or they always get their own way, the quality of the contact they have, of their relationships, is tense. There is little mutual pleasure and much irritation. All members of the family feel trapped in an unpleasant cycle. This can lead to despair and rage. A child who gets his way by bullying never feels anything is satisfying because it is not freely given. There are no gifts, only extortion. He may feel powerful, but not valued or loved. This is true in small ways and not just in dramatic circumstances such as Darren's. It is why the child who always cries 'I want' and gets it rarely feels he has enough.

Mr and Mrs M had been trying for a baby for many years. They were both well established in professional careers but felt their life was empty without a child. Finally, after a course of fertility treatment, Mrs M conceived, and Caroline was born, a healthy, pretty baby. Both parents were over the moon and very indulgent of their child. They wished to spare her any distress and rarely told her off. Mrs M gave up her job and spent all day with Caroline. By the time Caroline was four, Mrs M was the envy of many parents – she never got irritated or shouted, she had endless patience. She was seen as the 'best Mummy' by Caroline's friends. She cooked every meal, baked her own bread with Caroline, was always prepared to play, to organize craft activities, to sew dressing-up clothes. Everything Caroline wanted, she got. She hardly had to make a fuss and she would be indulged. However, instead of growing into a contented, cheerful child, Caroline often seemed sullen. When one activity was over, she would ask for another straight away. When her mother set up a game, she would change her mind and demand another. She seemed discontented. If she was unkind to another child, Mrs M would laugh it off, saying, 'Oh, Caroline, that wasn't very nice, I'm sure you didn't mean it.'

Caroline's friends got fed up with her, finding her spoilt, and she was not a popular child.

As Caroline always got what she wanted, by complaining or just demanding, she grew used to getting her own way. She had no practice at making compromises or waiting. When in the company of other children, she was at a loss. She did not know how to share nor how to play with them on an equal footing. Mrs M's wish to spare her any pain backfired. It left Caroline without the necessary skills to be with others who did not have just her needs as their priority.

Edward, two years old, is the only child of a single mother. He is very beautiful, with big brown eyes and a fetching smile. He thoroughly enjoys other children's company but gets easily over-excited, and grabs them and squeezes them, occasionally hurting and frequently frightening them. His mother gets embarrassed and apologizes for him but does not stop him from doing it again, or at least not in time. He is very much her precious baby and she is thrilled at his gregarious and outgoing nature. However, she appears to be blind to his aggression. Eventually he becomes quite unpopular at the mother-and-toddler group they attend, and other children are wary of him. Ms N is embarrassed and feels the others are unfair to her son. She is quick to defend him and to reprimand children she feels provoke him. Edward becomes more openly aggressive, even to Ms N.

This is a very common dynamic, often seen in groups of mothers and young children. Usually the situation is resolved by the mother slowly recognizing a side of her child that is less than perfect. On the whole parents can adapt to and accommodate naughtiness, jealousy and aggression. Unfortunately in Edward's case, there was so much invested in his being the dreamt-of child that his difficult behaviour was not sufficiently acknowledged. In wanting to keep her image of him as her good little boy, Ms N could not perceive other aspects of his character. In that way Edward could not feel secure that he was loved for who he truly was. He tried to communicate his anger and rivalry with other children. Because his feelings were not understood, he did not get help in managing them.

Over the years, he became more and more difficult, at home and at secondary school. His relationship with his mother became fraught. Eventually, their contact became increasingly negative. Edward believed he was seen as uncontrollable, which increased his anger. In addition he felt guilty for not being what his mother wanted. Ms N felt very disappointed, angry and unhappy. What had been impossible in the early years was to see Edward as a mixed bag, positive and negative, from the start. Blindness to certain feelings meant that he had to struggle with them alone, which he could not do satisfactorily.

We can also see from Ms N's point of view that he may have represented an ideal child. To recognize his aggression fractured the image she carried of him. Being alone, she also did not have the support or even the perspective of a different view. She, too, was deprived of a helpful person who could tolerate disturbing feelings and bear them with her. The early pattern of not saying no to Edward hindered his development and stopped him from finding ways to deal with the more difficult side of his personality. Those aspects of him therefore remained immature and got him into trouble as he grew older.

The Benefits of Boundaries

FEELING SAFE

We have seen how a child who dominates adults is in a very frightening position. If at the age of three or four you feel you are more powerful than those who look after you, how on earth can they protect you if the need arises?

From the child's perspective, limits may be restrictions and they may be infuriating, but they are also like gates, keeping things safe. There are other good reasons for limits. There are, of course, the obvious ones of physical safety – not allowing a child to play with dangerous objects such as plugs, fires, knives. Things get a little more complex when negotiating whether a child is allowed to walk by your side or has to hold hands when crossing the road. Then there are the numerous occasions, every day, when you need to set some gentle but firm limits which are not directly linked to safety but which help the child develop a sense of security.

After a morning of being at a mother-and-toddler group, Amita wants to carry on playing through lunch. Her mother says, 'No, it is now time to eat.' Amita shouts and stamps about and says she will not eat.

If Amita is allowed to eat on the hop, carrying her food around with her, she may at first feel triumphant; she may also feel Mum was not able to stand up to her, as we saw earlier with the bullies. She may feel Mum can't be bothered, and then seek other ways of grabbing her mother's attention. Concessions made for an easier life rarely prove effective. If her mother manages to be firm and help Amita with her crossness and Amita is able to eat and enjoy her food, everyone will come out a winner. They will have a sense of togetherness and achievement at having recovered from their conflict.

This could apply to all sorts of other situations, such as a child having to wait for something he wants, having to play alone for a while. In Amita's case the boundaries also help her realize her crossness or refusal are seen within the context of her as a whole child. Mum may be saying, 'I know you are angry and want to do something else now, but I also know that this is lunchtime and it is better for you to eat quietly. I don't mind that you are furious and I will not give in to your rage, but I will make sure you have what is right for you.' These are obviously not the words the mother would use, but the gist of her communication to Amita. This does not even need to be verbalized; it can just as easily be communicated by actions. Through her mother's stand, Amita gains a sense of feeling protected. She benefits from the reassurance that what is best for her is being served lunch despite her resistance. Feeling that somebody is prepared to put up with unpleasantness in your interest boosts security.

GROWING STRONG

The other important aspect of limits is that they help develop one's own resources. If somebody else does all the work, grants you your every whim, you become weaker and are increasingly unable to cope with frustration. The well-meaning parent who wishes to spare her child every pain could be preventing him from developing ways to deal with difficulties. Obviously here there is a judgement to make about what is bearable for the child and what is the difference between need and greed.

Children will integrate what they are learning about your limits, but at their own pace. You might well see a toddler purposefully spilling juice from his beaker on to the floor, saying, 'Naughty, no make mess.' At that point he is struggling with part of him that knows he is not meant to do this but at the same time cannot resist it! Learning to abide by rules takes time and is hard work, which must be appreciated.

Every boundary that is set is also an opportunity to develop. Having to eat when she wanted to play gives Amita a chance to resolve a conflict. If she succeeds, it begins a pattern of her believing she can get over difficulties. Mum insisting that different activities have a rhythm helps Amita know about structure, events having a beginning, a middle and an end. This will help her bear hard times and make use of that experience when she is enjoying something.

A child who wants attention, or a particular toy, or an activity, and has to wait or give it up is also learning to be flexible, to be patient, to think of alternatives, to be creative. These are all skills which are helpful in life. A child who has to play by himself because mother is busy may explore his surroundings, find a box and start to play a game with it. He may make it into a castle or a bed or a space-ship. He will use imaginative play to create for himself the company he wished for. A younger child will bang it, turn it over, put it on his head; like a little scientist he will find out all about the properties of the box. It is in the moment of frustration that we are given the chance to reach inward for resources, providing the 'no' is realistic, so that the child is not pushed to despair.

Fighting the Limits

We all know it is not easy to be the one told 'no'. If you say no to what the child wants, you have to be prepared to cope with the reaction.

TANTRUMS

The tantrum is a characteristic feature of toddlers' lives. They can get intense with rage and act as if they are truly falling apart, throwing themselves on the ground, limbs flying about. Our reaction tends to be of anger or anxiety that they will damage themselves. You may also be embarrassed by their lack of control. How it makes you feel is the main communication of a tantrum – you are meant to worry, to feel helpless, cruel, or whatever. This is how the child feels. The French talk of somebody being '*dans tous ses états*', that is, 'in all his states'. We speak of 'losing our temper', being 'beside ourselves', being 'all over the place', as if we really have lost something that belongs to us, a part of ourselves. Tantrums are a demonstration of losing a coherent sense of self and feeling fragmented.

These moments can be frightening to experience as well as to witness. When young children are upset, they tend to act rather than speak, their communication is through their behaviour. If an adult can count to ten and gather up the child and try to make him feel more together, less in bits of fury, the chances are he will calm down. The work of the adult is to remain calm and not get so flooded by the child's feeling that she is taken over by it and proceeds to have her equivalent of a tantrum too. Tantrums are not to do with reason, and often when witnessing one or being the cause of one, a part of us which knows what it feels like to fall apart gets stirred up. We want to stop the experience quickly and can easily get drawn into the conflict, rather than keeping the necessary distance from the child's state which is needed if we are to help him. It becomes easier to get cross and say, 'Just stop this nonsense', than to see the child as distressed and in need of calming and holding.

When children are young, we are still able to hold them physically, to gather them together till the wave has passed and help them recover. Some children may need this physical holding, others may be held by your voice or your patience or your letting them finish in their own time with you just being present.

PARENTS AS MONSTERS

Some children may act as if you have turned into the Wicked Witch of the West, and you may doubt yourself, wonder whether what you are doing is really cruel. You may have to remind yourself, for instance, that saying no to yet another video is not nasty and probably quite a good idea. My daughter coped with her anger with me by crying and sobbing, while in my arms, 'I want my mummy!' as if to say, 'Not you, you horrible person, my real mummy, the nice one!' It appears to be very hard, when they are little, that they should feel so mixed about the same person. If you think about it, this is something which stays with us all our lives, in less obvious forms; we often find it difficult to understand how the person we love can make us so exasperated at times. This is a common way of dealing with ambivalent feelings – to split them up rigidly, to think of one person as good and the other, usually the limit-setter, as bad. This can cause all sorts of trouble between couples or between parents and nannies and other care-givers, grandparents, teachers. As in fairy tales, splitting, in this sense, is a way of keeping things simple, of being justified in hating or loving with passion. Mixed feelings are much harder to deal with.

Understanding the process helps us think more clearly and avoid taking the criticism too personally; this in turn makes it easier to stay firm. If you can hold on to your image of yourself as doing the right thing for your child, you have more conviction. If you believe their view at the time, that you have become wicked or cruel, you may turn into the cruel person or feel paralysed by the image, as we saw with the example of Jack and his dad at the beginning of this chapter. When you feel a child turns you into a monster, this can be very upsetting. It makes it hard to think clearly. You need to take time to look at the situation objectively and examine what is really happening. Is their perception accurate? Sometimes it may be: you may recognize that you are being unduly harsh, and you can then modify your stance. At other times you may see that you are doing the right thing by being firm, even if they don't like it. You then have to be prepared to be unpopular.

ANGER

Limits often provoke anger, and we have to be able to face this. Anger is common to us all, yet we frequently attach guilt to it. However, it is normal and healthy to feel angry about certain things, and reassuring to children to know that parents feel it too. The difference is in what we do with the feeling. If parents can be angry and get over it, the child, too, will learn to manage his feelings in a positive way.

It is important that children are allowed to feel angry and to learn acceptable ways of expressing this. Young children vary enormously in what they will or won't make a stand about. In a nursery you may see a child who will protect a puzzle he is doing as if his life depended on it and another who will just wander off if somebody takes his pieces. Similarly, if hit by another, one child will scream and cause a major commotion, while a second will cry and go to find help, a third may hit back, and a fourth will just quietly withdraw on his own. You would hope to foster in your child a sturdy sense of himself so that he can feel anger at being badly treated. How he expresses that anger will predict how he, in turn, is reacted to.

In the example above, you might help the first child to put the hitting in perspective: was it a major trauma or a simple confrontation? For the last child, you might wish to encourage him to stand up for himself and not allow others to hurt him. Everybody in life feels angry at times; all families and social situations have conflicts. A great skill we all need to learn is how to manage conflict and strong emotions.

I would like to turn to a very different example, where ordinary anger and fear can go unnoticed because they are not clearly expressed, with the story of Alan, whom I saw following the death of his baby sister.

Alan was four years old and some months previously the family had lost their new-born baby, a few weeks after birth. Alan was a very bright and articulate little boy, always well behaved and thoughtful. He had been very kind to his parents and tolerant of their absence during his sister's illness. He was referred to me because he had become increasingly frightened, obsessed about time, particularly when he was due to be collected by his parents, and very anxious about losing things. His parents and his nursery were worried about how he would settle at the new school he was soon due to start. Although wonderfully supportive of him and understanding of the effect of his sister's death on him, they were at a loss as to how to help him.

In his play Alan was totally preoccupied with emergencies – ambulances, looking for a good hospital, fires and firemen, policemen and robbers. Although at first he was always in the role of the rescuer, slowly the figures changed and policemen became bad, firemen set the fires, and so on.

Of course we discussed the obvious level of his play being related to wanting to rescue his sister, to find a good hospital which could have made her better, and his despair that none of the rescuers had been able to keep her alive. This is the side of his feelings that everyone shared and recognized and was trying to help him (and themselves) with. However, I felt it was important to see and acknowledge that sometimes these good figures weren't so kind after all, and that sometimes in his play they were the trouble-makers. We were then able to talk about how he hadn't always felt kind about having a sister, and how these thoughts had left him feeling at some level responsible. This made him frightened, not just about what had happened but also about what might happen in the future – to his parents, for instance, if he was angry with them. It was hard for him to feel free to be upset or angry or to behave in anything other than a kind way.

Once this area was opened up, he became much more naturally mischievous, bossy and cheeky in his sessions with me and in his play with the toys. He behaved as any active boy his age would. I was able to talk about this with his parents and him. Although they had been grateful for his easy behaviour in the past, they were now able to help him to express his negative as well as his positive feelings. This also made it much easier for him to settle into the new group at school, with the hustle and bustle of boisterous children.

The expression of anger has to be tolerated by others in order for it not to feel unbearable or, at worst, deadly. Unless anger or rage can be voiced, it becomes difficult and at times nearly impossible to manage these extreme feelings. The child has no way of learning to control his aggressive emotions unless he is able to experience them himself. This is the only way he can know how strong they are. If he cannot speak out his fury and rage, or act out some degree of how he feels, he may imagine his powers of destruction to be far greater than they are. He may be in for a shock, for instance, when faced with a little bully at playgroup. Depending on his experience and what he's made of it, he may behave as the very good boy who never fights and then get clobbered; he may think he is invincible and be surprised to find the other boy is much stronger; he may be uncontrolled and really hit much harder than he thought he could.

There should be a legitimate space for feeling angry, thinking angry thoughts. The expression of anger and how acceptable it is varies enormously, culturally, even within different families. Slamming doors, shouting, breaking things, may be the norm in one household and considered absolutely extreme in another. The child has to have a taste of his own anger within his family and then compare this to what is allowed elsewhere.

AGGRESSION

One way children have of dealing with anger is to be aggressive. Aggression is often the flip side of fear. Young children, particularly, act very fast on their feelings; they need an adult to give them a model of thinking before acting. As we saw with Paul earlier, his aggression was born out of fear. He hardly dared have the feeling and usually acted before he even knew what he felt. Through Paul and other deprived children, I was made deeply aware of the need to talk and think about how such children feel and to try to put these feelings into words so that they can develop a way of making a gap, a distance, between the feeling and the action. Although Paul is an extreme example, all children may share some of his experience in a small way.

Children are often aggressive when they are frightened or feel under threat, whether because they have been told off, or because they cannot do what they are asked, or because they are in fact bullied by others, children or adults. One way of not being scared is to be like the scary ones, to become the aggressor. Being little can be frightening – everyone else seems to be able to do things better, to have more power, to be bigger and stronger.

A lovely book which illustrates this is Maurice Sendak's *Where the Wild Things Are*. It tells of a rather mischievous little boy, Max, who gets sent to bed without his supper for his naughtiness. As he sits in his room, it turns into a wild forest, and he then travels to the land where the wild things live.

I think the story illustrates so many aspects of the 'wild thing' in young children. Max's hurt at being told off makes him retreat into a fantastic place of scary figures, and parents *can* be frightening when angry. The journey takes ages, as time out does, when you have been isolated. Finally, he finds refuge with the wild things. How else to deal with them than to become the king of them all? But once the wildness and rebellion has been indulged, it is really a lonely place to be. Max then feels the monsters are the naughty ones and, as his mother did with him, he sends them off to bed without their supper. Luckily for Max, he also remembers and holds on to the image of his mother as 'good enough' and so 'from far away across the world he smelled good things to eat'. We can tell he is a much-loved child because, in the happy ending, he finds his way home to a warm supper – the idea of a mother who still loves him in spite of the wildness.

Unfortunately for Paul and many other children I have worked with, the way home is not always easy to find, and the wild things in the forest are often safer than those at home and their call to stay is very appealing. To many children the wild place becomes a permanent place rather than an occasional journey. There are alarming pieces of research showing that 'childhood trauma leads to violent behaviour not only in childhood and adolescence, but also in adult life'. In working with young children, we are in a good position to break the cycle and help them to have a different experience.

We must not forget that aggression can also be a positive force. It can be the emotional equivalent to muscle tone. It is often aggression which gives determination, a pull forwards. It is what we do with our aggression which turns it to constructive or destructive use.

HATE AND LOVE

With time we learn that a thought is not the same as an action. Also, with more thought before action, we can start to make a choice about whether we want to act. My little Paul learned to say, 'I want to smash the window,' rather than doing it. The other big advantage of thinking before we act is that we learn about the consequences of our actions.

As parents, we have to be able to take the fury and the rage when we make a stand about things and say no to our children. We have to learn how to make rules about living with others in the hope of enabling the child to have consideration for others.

Jeremy, a delightful and articulate boy I saw between the ages of two-and-a-half and five, was referred to me because his behaviour was very wild. His rather quiet middle-class parents found it hard to control him and were overwhelmed by his excited, violent and volatile behaviour. The precipitating 'last straw' was his clouting his father on the head with a hammer, when Father was playing the piano and Jeremy wanted his attention. The following excerpt is from a session, after a holiday, when he was four:

Jeremy came storming in. He started off roaring, stomping around the room, threatening to 'eat you all up'. He could not settle and rushed around, cutting paper, scribbling, playing at making the water in the sink gush, and so on. He checked around the room and, in a booming voice reminiscent of Daddy Bear in the *Goldilocks* story, he said, 'Who's been here?' I talk about the holiday making him angry and how he is being like a giant checking everything out, and wondering who has been with me while he was not. He starts to pull the furniture out of the doll's house roughly and says, 'It's dead, you can see it.' He pulls bits apart and starts to bite them. I say he is so angry, he feels like a lion roaring and biting bits off and eating them up and making things dead. He says very quietly that when he cries, he wants Mrs Phillips. I say he really found the holiday hard – we missed our time together. He says, 'If I eat you up, then you won't be there, will you?' I say that when you love people, like Mummy or me, you feel things very strongly and you are worried you will hurt them and lose them.

The point here is that it is irrelevant whether or not he could eat me all up; the truth emotionally is that he was so angry, he wanted to do so. As we saw before, in fantasy anything is possible. Also, he may have had the idea that if you eat someone up, they are inside you and cannot ever go away again. In our time here, he was able to think beyond the fury to what followed – to the fact that he also loved me. It is in that ambivalent state that we start to see the other fully, with good bits and bad, and make a choice on balance as to how we feel about them, as a whole person. It is also through being able to stay with the fury and recognize it that thought can begin and with it a capacity to appreciate the other, the beginnings of gratitude. Jeremy and I had many opportunities to explore strong feelings and their impact. This was as true of passionate, positive feelings as of fury and hatred. I am sure I learnt as much from his honesty and straightforwardness as he did from me. No prizes for guessing how very fond of him I was!

Everyday Limits

I have tried to show that all behaviour has meaning, that it occurs within the context of relationships, and that there are therefore many sides to the same story. We have looked at the ideas of dealing with mixed inner feelings and at how our past experience can colour the present. We have seen some of the consequences of saying or not saying no, and at the reactions we are likely to encounter. I would now like to discuss setting limits for some of the more common difficulties and behaviour we come across with young children.

SEPARATION

A clear time when we say no to a child comes at the moment of separation. He wants to stay with you but has to be with others. One regular complaint from families is that their child will not separate easily, he clings and whines. The dilemma is not just the worry about how the child will get on with others, but that if you don't negotiate the separations well, more often than not your contact and togetherness is not very pleasant. The child is clinging to you while you are trying to push him away. Here again, we should look at what the behaviour means to each person. It is important to listen and hear the story. Separation is always a two-way process.

One of the major early separations for young children is being left with a granny, au-pair, nanny, or going to nursery or to a child-minder. Anyone who visits a nursery will see at least one parent who is very concerned about letting the child go, although appearing to push him away. You may notice that when the child has said good-bye once and moved off, albeit a tad reluctantly, ten good-byes are said by the parent, indicating that one was not enough and that the process is painful for the parent too. By the end of the scene, the child may well be in tears, having picked up and expanded on the parent's feelings. This is the reason some nurseries do not allow parents beyond the front door. That, of course, does not solve the separation problem for parent or child, but it may make life easier for the staff on a short-term basis.

How time away from the parents is presented will determine how the child experiences the separation. Does the mother trust the person the child is left with? Are her own memories of being away from home good? Does she feel that the child is being left for his needs or hers, for instance if she is going to work? All these factors will have an effect on how she says good-bye. A good-bye said with confidence will give the child the prospect that the next few hours will be good.

There is also the crucial issue, so often raised, of whether you say good-bye or think you can disappear unnoticed. Many adults fool themselves that young children are unaware of their surroundings – out of sight is out of mind. Having worked with under-fives for many years, I can guarantee that this could not be further from the truth. What is true, however, is that if children are told of an impending departure, they have a chance to object and fuss. Parents need to recognize and accept these feelings while sticking to their plan to leave.

Another alternative which can cause difficult feelings is that the child may be happy to let them go! This can be very painful for the loving parent who is sensitive to how much she will miss her child.

What do you do with a child who screams, hangs on for dear life and looks as if he will die if you leave? As with the examples above, you need to think whether these feelings really reflect the situation. If you leave anyway, in a positive manner, with appropriate confidence that he is in good hands, you are reinforcing the idea that he will be OK without you, that there are other people in the world who can also look after him. If you do not leave, you are in effect agreeing that only you can look after him, that the world at large is not safe. Of course the leaving has to be done thoughtfully. The time he can manage without you also needs to be considered. It may have to be a slow process, but it has to be begun if your child is to taste other treats than those you offer.

SLEEP

Dilys Daws, a child psychotherapist, has written very clearly about the impact of parents' feelings about separation and how these are communicated to children. In her book *Through the Night*, she links sleeping difficulties to parents' ambivalent feelings about separation.

Many families of today require bigger and bigger beds to accommodate the children who creep in next to their parents. It is very much a difficulty of our time. There are conflicting ideas about the benefits or disadvantages of letting children into the parental bed. It must also be said that it is rare for an older child, adolescent or adult to sleep with his parents! On the whole, therefore, we can think about the issue mainly in relation to children between two and six or seven. Why do we find it hard to be firm about our children going to bed in the first place and then to say no to them getting into bed with us? What interferes? As always, the answer is not simple. It could be many things.

In chapter 1 we discussed the fear of separation. Feelings of loss are often stirred up by sleeping; in an extreme form there is a fear of death. Tennyson wrote of sleep as 'Death's twin brother', and most parents, before going to bed, will go to check their child is all right, still breathing.

In letting yourself drift into sleep, you are also entering a time or space you do not have much control over. There is a sense of isolation in sleep – for some it is a restful haven, of privacy, calm and nice dreams. For others it is a tempestuous world of nightmares. For most it is a mixture. But we do not know in advance what our sleep will be like. Common language makes it sound like a journey: we wish each other good-night, pleasant dreams, we say, 'See you in the morning' – all markers of leaving and returning.

The parent's view of sleep will have an impact on what the child anticipates in that twilight zone between wakefulness and sleep. A parent who starts off the night leaving a light on, in case the child is alarmed by the dark, is already anticipating that darkness is worrying rather than restful. The same goes for open versus closed doors, quiet versus noise. How sleep is viewed and catered for communicates an image of what it is. There is a beautiful children's story *The Owl who was Afraid of the Dark*, which illustrates how a little owl gets over his fear of the dark by speaking to creatures who love the dark. If you feel an empty bed is a treat, you are more likely to be firm with your child that he should sleep alone. If, however, you feel it is lonely, you will imagine that is why he wants to be with you and will allow him in. In fact, he may have a totally different reason for coming into your bed or not wanting to go to sleep. You will need to listen to him rather than assume he is like you.

There are many other explanations for why parents allow children to stay up very late or to sleep in their bed. A mother or father who is at work all day may feel the only time with their child is in the evening. They have a feeling of closeness snuggled up to him at night, it makes them feel in touch. The occasional night all together may be very satisfying all round. It is important, though, to know whose needs or worries are being met. A single mother I worked with would often have her little girl in bed with her because she herself was frightened at night. Although she clearly knew her two-year-old could not protect her, she felt safer with her in bed.

Other causes can be due to marital problems. A child in the bed, ostensibly so that both parents can see to his wishes, may hide a distance they wish to put between them. In an unhappy couple, the child in the middle can help both parents avoid their sense of loneliness while acting as a barrier between them. This sort of situation is not helpful to the child. He would sense the parents' need for him and find it hard to protect his own space. Some children may worry about the hostility between their parents and want to be there to ensure nothing bad happens. The parents, too, might be grateful for the safety net he provides. Other children may feel the rift and want to fill the gap, to be close to one parent and push the other one away. All these questions and more may have to be considered if you find that your child is always in your bed.

Allowing a child to be with you all night, every night, is not helpful to him. It stops him developing a sense of himself on his own. A child who is afraid at night and who is then taken into bed regularly does not develop strategies to cope. He is thus always vulnerable. Night after night, he will be afraid and call for you, rather than learning to hide his head under the cover, or sing to himself, or listen to a tape. If he believes there are crocodiles under the bed or goblins in the cupboard and you remove him from the room, even if your words deny their existence, your actions suggest he is better off out of the room. He does not have the experience that if he stays, he will find out there are no beasts in the bedroom. On the other hand, if he has to try out methods of dealing with his fear, over time and with the help of various strategies, the fear dissipates and is conquered. The child grows in strength and confidence in himself and in his resilience. In his science-fiction epic novel *Dune*, Frank Herbert creates a 'Litany against Fear' which I think describes well how we successfully conquer it:

> I must not fear. Fear is the mind-killer. Fear is the little-death that brings total obliteration. I will face my fear. I will permit it to pass over me and through me. And when it has gone past I will turn the inner eye to see its path. Where the fear has gone there will be nothing. Only I will remain.

With experiences like fear, for a young child all the reassurance and talking in the world will not convince him as much as feeling it, surviving the emotion and coming out of it.

A child may only remember his fear at the time of going to bed, but the parent can think about it during the day. This offers an opportunity to talk to the child about his fear and think together of ways to tackle it before it has turned to panic. Then when the time comes, the child and parent may have found ways to help him deal with it – plans and strategies for him to call on. This could involve having a special toy with him, hiding under the sheets, listening to music or a story, leaving a night-light on or any number of solutions.

FOOD

Many young children get into terrible battles with their parents at mealtimes. Whether it is a question of what, where and how they eat, it is often a rough ride. A mother, particularly, can be sensitive to reactions to the food she provides. She may experience her child's refusal to eat as a rejection of her all together. Sometimes there is so much emotion stirred up by meals that it may put the child off. He may find the atmosphere indigestible, rather than the food. We have all experienced a knot in our stomachs when we are in a stressful situation. Young children, especially before they have a firm control over language, are very sensitive to emotional climate. So a child who rejects chicken soup may just not like it, but if he is treated as if this means he hates his mother, he will find it hard to eat anything. Many Jewish jokes are based on this interaction, common among mothers and children. Alternatively he may comply and stop discriminating for himself between what he likes and dislikes; he will fit in, for the sake of peace.

On the other hand a child who refuses to try anything new or acts as if all food is bad may need a mother who can translate this reluctance for him. As we have seen all along, how we react helps to give the child an idea of the world around him. A mother who allows the child to be extremely picky is, in effect, agreeing with his view that there is not much that is good to eat. A mother who cannot say no to her child's demands for the same meal every day, or allows him to refuse to eat most things, can end up tyrannized into accepting the child's orders. Once she starts changing all her usual ways and catering to his every whim, she becomes over-preoccupied with what he will or won't eat. She becomes tentative, which in turn makes him suspicious about what is on offer. She may also feel guilty that he is not having a healthy diet. Food becomes a source of displeasure. What a loss all round! She could make certain rules: that he has to try something, or that he can choose a limited number of foods he doesn't want to eat. For instance, they might reach a compromise where spinach, beans and cabbage are off the menu, but she will expect him to eat other vegetables. Such rules respect the fact that he has his own taste, while allowing the mother to maintain her position that many things in life are delicious. As we have seen throughout this chapter, it is important for the mother to feel confident that what she is offering is good; she will then present this image to the child, who more often than not will enjoy his meals.

WAITING

Young children, as we know, live intensely in the here and now. They have a very subjective sense of time. Waiting is difficult for them; they want instant gratification. Some of this is quite physical

– we have all seen a hungry child being fussy, irritable and irritating. The transformation after he has eaten seems almost miraculous: he is cheerful and friendly. Similarly the child who is going down with an illness may first show grumpy behaviour; it is only when the illness comes out that we realize that was what the grouchiness was about. (After experiencing this over and over again, I decided that poor behaviour is quite a good diagnostic tool of physical illness, almost a symptom!)

When a child has not got what he wants, his feeling that waiting hurts him is partly based in reality, in his experience. However, he also needs to know that waiting can be OK – that he will survive it and the feelings aroused in him. Often his reaction is such that the mother also believes he cannot bear to wait, she will stop whatever she is doing to see to him. It is a frequent complaint of friends that they cannot have a decent conversation with a mother because she will let the child interrupt constantly. The repeated experience of waiting for a time that is tolerable will give the child practice and confidence that he can manage on his own.

Mothers who feel cruel making a child wait may be so closely identified with him that it is their own infantile side they are indulging. If *you* find it quite painful, it is hard to give your child a different picture. Your struggle adds to theirs. Again, it is important to disentangle whose feelings are dominant.

Guilt also hinders our capacity to help children wait. Many a mother has felt guilty for strictly scolding her young child and then finding, a day later, that he has got the most awful cold. At worst we can feel our scolding made him ill. There is a sense that saying no, setting a limit, is dangerous. However, we have seen that not doing so is actually harmful. The child who cannot wait is at the mercy of his intense emotions and can feel very unhappy. Giving him a limit can help keep those feelings within bounds. Otherwise he may feel full of wildness which never gets tamed, just as we saw with the children who behave as if they are omnipotent.

DESTRUCTIVE BEHAVIOUR

Whatever the circumstances, it is important for the child to learn not to hurt others, and if he does, to learn to repair the damage done. Children who are allowed to be destructive become very alarmed, both about what they have done and about what you might do back to them. Paul, the violent little boy I discussed earlier, was once taken for an X-ray. As the machine came towards him to take the picture, he screamed and yelled, 'Don't shoot me!' clearly petrified that the hour of retribution had come. Children who break things repeatedly and smash up whatever frustrates them may grow up to feel guilty at first and eventually may despair that anything good will ever survive. They end up feeling the world is broken and hopelessly beyond repair. It is a relief when somebody stops you from hurting others; it means that they would also be prepared to protect you. When a child is in a whirlwind of fury, he may feel out of control. Here again, to be stopped firmly is a sign of care – a willingness to face their wrath with their welfare in mind.

We have seen reports in the news of young children brought up on the fringes of society where the impact of what they do seems to matter not one bit. No one is looking after them, making sure they are not being destructive and in trouble. These children do not feel what they do makes any difference. They adopt a 'who cares' attitude. Letting them behave in this way is allowing them to build themselves a bleak world.

MANNERS

One of the new tasks of children between two and five is to master the niceties of being in company. Whereas on the whole we do not expect much of babies in this area, the toddler and young child has demands put on him. To be exact, the baby has had a little taste of this – being asked to smile or to wave good-bye, to be friendly with people other than his parents, and so on. But at this later age, a child may be told off for not complying.

Here too we are faced with the task of helping the child fit in without crushing his sense of himself. The balancing act, between what we think is appropriate and what he can deliver, is ever present. It is crucial for children to learn to behave well in company for the simple reason that if they do not, nobody will want to be with them.

We may think of childhood as a wonderful time of freedom and be loath to restrict them. Alternatively, we may think children are just little adults, be exasperated when they do not behave as such, and bully them into conformity. Our philosophical stand, on the advantages of childhood and the necessities of being a part of society, will have a great impact on how we enforce our limits. Our own struggles to be individual and yet part of the group will enter the frame. We must therefore act with respect to the emerging personal style of the child, while giving him the tools to get on well in life. Mostly, a consistent and repetitive approach rather than a heavy-handed one works best.

It is important that we, as parents, value being treated well. A mother who allows her child to be very rude and inconsiderate is saying that this is a fair way to treat her. She is giving in to an abusive aspect of the child which, as we have seen, is not healthy for him. She is also setting an example: he will not have the force to stand up for himself if people behave badly towards him. For mothers who do not have a strong sense of self-esteem, or who base their view of themselves on always being available, it will be harder to stand firm. We must remember that what makes an impact on our children is not just how we behave towards them, but how we allow them to behave towards us. Manners and social conventions are not just superficial; they are originally and fundamentally about having relationships. Growing up involves a degree of being tamed. A very touching passage in the story of *The Little Prince*, by Antoine de Saint-Exupéry, is the fox's request to be tamed. The little prince wants to play with a fox he has just met. The fox answers that he cannot play because he is not tame. The little prince is puzzled and so the fox explains that 'tame' means 'to establish ties'. He goes on:

> 'If you tame me, it will be as if the sun came to shine on my life. I shall know the sound of a step that will be different from all the others. Other steps send me hurrying back underneath the ground. Yours will call me, like music, out of my burrow. And then look: you see the grain-fields down yonder? I do not eat bread. Wheat is of no use to me. The wheat fields have nothing to say to me. And that is sad. But you have hair that is the colour of gold. Think how wonderful that will be when you have tamed me! The grain, which is also golden, will bring me back the thought of you. And I shall love to listen to the wind in the wheat . . .'

A mother who helps her child build relationships in this way will always be special to him. She will have earned and deserved his gratitude.

Our Place within the Family

A child may be born into a family with siblings or he may be an only child. Between the ages of two and five is frequently when a new baby is born, which raises the question of our place in the family. Changes may have to be faced and sharing is likely to be hard.

In order to share, the child has to start from a position of security; he has to know he is deeply cherished. The advent of a new baby may shake his faith in this and the parents must reaffirm their bond to him.

I return to the story of the Little Prince, who loves a particular rose and tells the other roses:

> 'You are beautiful, but you are empty . . . One could not die for you. To be sure, an ordinary passer-by would think that my rose looked just like you – the rose that belongs to me. But in

herself alone she is more important than all the hundreds of you other roses: because it is she that I have watered; because it is she that I have put under the glass globe; because it is she that I have sheltered behind the screen; because it is for her that I have killed the caterpillars (except the two or three that we saved to become butterflies); because it is she that I have listened to, when she grumbled, or boasted, or even sometimes when she said nothing. Because she is *my* rose.'

MAKING ROOM FOR SIBLINGS

Research has shown that first children of this age group react particularly strongly to the birth of a baby. They are typically hostile and rivalrous with the baby, furious with and possessive of the mother. Their behaviour becomes difficult, demanding and often more infantile. They may want a dummy, suck their thumb, wet the bed. As with angry feelings, they must be allowed to express all this. It is also crucial that they be reassured that they are loved, in spite of being full of negative feelings. This gives them confidence in their position with you, and it helps them accept their feelings and integrate contradictory emotions. It may be a shock to them that the baby is here to stay.

A family I worked with in a Special Care Baby Unit were amazed at how well their three-year-old, Suzy, got on with the new baby. She never fussed about visiting or made herself a nuisance on the ward. When her mother was discharged home, Suzy was keen to come with her to see the baby every day and curious to watch her mother feeding him and changing his nappy. When the baby was well enough to go home, she was very excited and thrilled. However, at the end of the first day, she asked her mother when they were taking him back to the hospital. She had certainly not expected him to come and actually live with them!

Many a family has a similar story about the older child wanting to give the baby away or send it back to the shop, or whatever. To say no to their wish to get rid of the baby, to be firm that he is now part of your life together, is not easy. Many mothers identify with their older child – they may remember feeling ousted by their little brother. If they were the young one themselves, they may feel guilty for having taken so much of their mother's attention, and over-compensate by paying extra attention to the older child, who represents their older sibling whom they feel they have deprived. Like the older child, they too may feel invaded by this new, demanding and time-consuming little creature. On the other hand, they may feel very close to the baby, as if they are one, and resent the demands of their older child. All these thoughts and feelings will impinge on how they react. Once more, it is important to observe the child, to try to see his point of view and help him with his struggles, not assume they are the same as yours. At this time, the help of your partner or your own mother, friends and other supportive people is essential. It is very hard and painful to deal with such raw and conflicting demands all by yourself, without feeling overwhelmed.

The ultimate aim in setting boundaries in this circumstance is to ensure everyone has a legitimate place in the family, to try to permit all the mixed emotions a voice, without them actually being acted out. It does not help anybody to be destructive. Helping the child make room for others in his life prepares him for sharing in other contexts, especially the next stage of starting school.

WHEN SAYING NO IS ESPECIALLY HARD

There are some children to whom parents find it particularly difficult to say no. Families in which there have been fertility problems, and adopted, fostered or handicapped children all pose specific dilemmas.

It may be very hard to be firm with a child you have waited years for. During the time of yearning and repeated disappointment, an image builds up of the wished-for child. This can cloud our vision of the real child we end up with. It can make us blind to the child's negative side or hyper-sensitive to it. Either way, our reaction is likely to be slightly skewed. It is important to normalize the child's experience as far as possible. *He* has not waited years to be with you and is not part of the history that makes you see him in a special light.

With adopted and fostered children, you inherit an unknown past. Even if facts are available to you, there will always be a mystery about aspects of his personality. It is helpful to the child to feel he is entering a home which is clear about its rules. This gives a supportive structure to counterbalance possible past upheavals. Being firm about how you do things also gives the child a sense of belonging. It is reaffirming that now he is part of *your* family.

If you know the child has had a traumatic history, you need to hold on to the idea that this can be repaired. Children will expect to be treated as they have been in the past, and frequently make you feel as if you are like the figures they have been with. Deprived and abused children are notoriously difficult to look after. They behave to an exaggerated degree as little Jack did with his father at the beginning of the chapter. They may also be like Paul and push you to the limit. Separation will be a particularly poignant issue between you. When a child has suffered, our natural tendency is to spare them any further pain. It is important for such children to be cared for unconditionally, but it is also crucial to remember that caring is not indulgence. They are in desperate need of an adult who is prepared to put up with their resistance and their fury, and still stand firm. Saying no at appropriate times is very helpful to them.

If your child is handicapped in some way, you may have very mixed feelings about setting limits. It may seem cruel, too exacting and unmanageable. Here, as always, observing your child and seeing what he can handle is paramount. If you act as if he is too fragile to be told 'no', you are reinforcing the handicapped aspects of his personality. If you try to stretch him, you are giving him hope and a belief that he can manage. Of course this needs to be realistic and not lead him to despair. Especially in the early years, a way of finding out any child's abilities is by trial and error. One should not jump to conclusions about handicap without giving things a go.

Four-year-old Sam was watching his friends trying to hop on one foot. He was sitting, jigging up and down on his chair. His mother found it terribly painful watching him get excited by a movement he could not master. He indicated he wanted to join in. She was torn between letting him try yet anticipating his failure, and sparing him the pain she foresaw by distracting him with another activity. He seemed so keen she decided to take him down on to the floor and help him try to hop. With her help, he extended his legs, holding on to her hands and lifted his left foot up. He was thrilled with his achievement, as was his mother. He was satisfied that he had joined in and felt duly rewarded for his efforts.

Here we can see that the mother was prepared to face the sorrow that her son was not as able as the others in order to let him have an experience. From it came an insight that although Sam may not have been as agile as his peers, his determination and resilience were stronger than theirs.

A very understandable tendency with a handicapped child is to try to cram in as much as possible in the early years, because of the fear that development might suddenly be arrested. We know that high expectations can give a child hope and energy. But it is also important to see the whole child. You may need to say no to your wish to surmount the handicap at all costs. If you do everything in relation to the handicap, you lose sight of the individual. Sam has a handicap but he is not just a handicapped child. He is also Sam, with his own personality. We have seen that children get a picture of who they are by looking into our eyes, seeing themselves reflected. In our interaction with handicapped children, this is of paramount importance. The sorrow and pain will inevitably be there, the mourning at having lost the child we dreamt of. However it is our task also to make space for the child we have, to see him in the round, and not just his handicap. This will make it easier to apply all the principles discussed in this chapter with regard to saying no and setting limits, and help the child to have an ordinary life.

HELP AND SUPPORT

As we have seen, under-fives are passionate, fiery beings. They tend to swing to extremes, in their behaviour and in their perceptions. This is hard to deal with single-handed. The main carer will need support, somebody to dilute the intensity and to help bear and think about what is going on. It is hard to resist a whirlwind on one's own. Toddlers often rope you into behaving just like them; it helps if somebody can remain on an adult level.

This is also an age when the child becomes aware of hierarchy in a family, of his place within the whole. Knowing that he is not a partner to his parent but a child is a relief and a comfort, even if he complains about it. The strength of the couple will give him a secure base. The couple need not be his two parents, although it is usually so; it could be any two adults who come together to think and care for him. At times the other half may even be someone employed to look after him. Being part of a triangle and no longer part of the intimate mother-baby couple is a first step in learning to go out into the world. He has to learn to negotiate more relationships and to understand the idea of relationships which are not centred on him. He will also be in touch with the possibility of future babies, be these real siblings, or projects, activities and conversations which engross his parents and do not include him. These are all small and essential parts of healthy development.

Summary

A feature of the years between two and five is that the world is experienced as magical. Fantasy and reality are not yet properly separated and defined. The child's circle is very much a microcosm, based around the family and small groups. The issues that arise for him and his parents are often about how to deal with strong feelings, how to adapt to required behaviour, how to manage himself. Parents are still very close and, on the whole, act as an intermediary between the child and the outside world. This closeness helps them to understand him, but at times the differences between the child's perception and his parents' can become blurred. Some untangling needs to be done, observing the child's communications rather than assuming he is like you. Saying no, setting limits, will help the child have a model of how to cope when he feels overwhelmed, he will feel secure in his place in the family and begin to develop resources of his own.

Some of this development may be painful for parent and child but the rewards are great. In later years, the child forges ahead and at times is much on his own. What will serve him then is not the constant presence of his parents, but what he has been able to digest of his experience, what he keeps inside himself of what has been offered.

My final quote from *The Little Prince*:

> So the little prince tamed the fox. And when the hour of his departure drew near –
>
> 'Ah,' said the fox, 'I shall cry.'
>
> 'It is your own fault,' said the little prince. 'I never wished you any sort of harm; but you wanted me to tame you . . .'
>
> 'Yes, that is so,' said the fox.
>
> 'But now you are going to cry!' said the little prince.
>
> 'Yes, that is so,' said the fox.
>
> 'Then it has done you no good at all!'
>
> 'It has done me good', said the fox, 'because of the colour of the wheat fields.'

3

The Primary School Years

> Now this is the very first day of school (. . .) your headmistress is Miss Trunchbull. Let me for your own good tell you something about Miss Trunchbull. She insists upon strict discipline throughout the school, and if you take my advice you will do your very best to behave yourselves in her presence. Never argue with her. Never answer her back. Always do as she says. If you get on the wrong side of Miss Trunchbull she can liquidise you like a carrot in a kitchen blender.
>
> ROALD DAHL, *Matilda*

A Whole New World

Around the age of five every child has to make the transition from home to school. Some may have attended playgroups, nurseries, or been in a smaller group at a child-minder's, and others may have remained at home with their mothers or fathers. Whatever their past experience, starting school will be new. More and different things are expected of them and they have to adapt to the rules which serve the majority and may not be so attuned to their individual needs. The focus shifts from acquiring mostly social skills to educational ones. Children, in this phase, have an enormous amount to master. Their response to rules, regulations and manners – all areas related to boundaries – is of paramount importance. How they react to 'no' will have a major impact on their capacity to settle, make friends and learn at school.

One of the main worries when a child starts 'proper' school is how he will cope with the teacher, in the group and with the demands put upon him. In Chapter 2 we looked at the problems which can arise with separation and these are likely to recur with the beginning of school. The child may feel daunted by the prospect of long days away from his parents, he may feel very lost in a crowd without the security of a mother or father as an intermediary to monitor his contacts. He may be unsure about his relationship to his teacher – is she like a parent, an aunt, a policewoman, a helper? He will feel a mixture of being quite grown up, going to 'big school', and at the same time little, unsure of the routine, of what is expected of him.

It is a time of great excitement as well as of worry. As we saw with babies, there is not just a weaning from, but a weaning to. Most children take starting school in their stride and embrace their new identity as school children with gusto. They love the symbols of their new status – the uniform if there is one, pencil-case, packed lunch, homework diary, exercise book, and so on. My nephew insisted on taking his satchel to bed with him for the whole first week of school!

Parents may not take to the experience with such enthusiasm. On the whole, they tend to be rather anxious; some worries are based on realistic concerns about their children and how they will deal with a new environment. Other anxieties may be linked to their own memories of school and the general feelings stirred up by change and separation. They are, after all, handing over their child, for the greater part of the day, to people who will contribute to the kind of person their child becomes.

When my younger daughter started primary school, a group of local mothers got together with the children the day before school started, so that the children could meet, make friends and start school with at least a few familiar faces. We also felt it would help if some of us knew each other a little. The idea was to try to make easier something we already saw as potentially difficult. We met; the children all disappeared and played happily with each other, while we chatted. As might be expected, we turned quite quickly to memories of when we ourselves or our older children started school. I was struck by how, very soon, our conversation turned to painful separations, and eventually we were discussing mostly hospital admissions and awful memories. Whereas the children were able to enjoy each other's company and the thrill of meeting new people, we, as adults, seemed to be preoccupied by disturbing and frightening thoughts. We realized this as we talked and it made us much more in touch with how hard it was for us to let go of our little ones and how worried we were about whether they would be all right. We were also quite surprised and shocked at the extreme nature of the events which came to our minds. None of us had been aware of how anxious we were.

Another interesting aspect of this get-together was that one of the mothers spoke of wishing to change her name, from Glynis to Gina. She had always disliked her first name and thought this would be a great moment to start with a new one, in a group of people who did not know her. This highlights the opportunities opened up by change. All of us have the chance to make a fresh start with strangers; we are not bound by what people already know of us. The children are being seen without their history, as it were, and have to make their own impact on their peers and their teachers. For a child who has a difficult time at home, or whose relationships in the family are fraught, it opens up the possibility of a new, more positive image. Naughty little Billy who is always relied upon to make a fool of himself may discover he can be quite serious. Quiet Gemma may find she enjoys the excitement of company and become a very animated girl in the playground. So our meeting brought to the surface the mixture of anxiety and opportunity that all new situations offer.

All these feelings will shape how we present school to our children. They will affect how we negotiate the stands we take on how the child behaves and what we hear about school.

Rules at School and at Home

Until now, the child has had a relatively small circle of people to deal with; he has grown accustomed to them and how they behave. He is familiar with certain expectations and tasks. He will have started and may even have mastered a bit of reading and basic maths. He will have learnt to get on reasonably well in small groups. However, with entry into primary school, the expectations rise sharply. Although most schools do think of the first year as a reception class, one in which the transition is carefully thought about, the change is still quite dramatic for most children. Even the simple routine of lining up to go out for break can cause distress. There is always someone who wants to be first in line, and inevitably some jostling and pushing. The whole issue of one's place in the hierarchy, the pecking order, is stirred up. Once in the playground, there may be many dramas acted out. The children have been sitting quietly during class and often let off steam. There can be feelings of great loneliness if one is not part of the gang, and for some it is very difficult to overcome shyness. Jane may think nothing of naturally joining in the game of 'It' which looks so exciting, but Craig feels he can't play with the others unless they ask him. He may wait all break, while the others don't even notice he is alone. Before, an adult might have spotted Craig and tried to involve him. At school individuals are harder to see. It is the general group which makes the greater impact.

The child has to become more self-reliant. He also has to gain more self-control. A pre-school playgroup or nursery is likely to have many activities going on at the same time and a certain amount of noise. Children will have a fair degree of freedom of movement. They may go from painting to working at the Lego table, then be taken for reading time as a small group. They often sing to themselves and jiggle on their chairs while working. Now in class they have to sit relatively still, be quiet, listen and concentrate. This is quite a tall order. It is easy to forget the amount of energy it takes to fit into this new setting.

The issue of rules is very prominent. Some will be explicit and spelled out, such as no running in the corridors. Others may be implicit and harder for the child to grasp. For instance, he may be able to speak to his neighbour but be told off for calling out across the classroom. He may have been used to using whatever pencils or pieces of equipment were around, but now each child has his own and he has to make sure he brings his in and looks after them. Whereas adults are used to telling toddlers and very young children what to do, the primary-age child is expected to exercise a great degree of self-discipline. He needs to have taken in the boundaries and set them for himself.

The main purpose of schooling is to learn. The child has to turn his attention and vigour to a specific task. His past experience will have a major impact on how he faces this expectation. Eva Holmes, an educational psychologist who did some research with very deprived children in day and residential care, writes:

> It is very likely that most infant-school children, brought up at home, assume that their teacher, 'my teacher', is personally interested in them, and is talking to them personally when she addresses a class; the presence of other children is probably somehow irrelevant. Children brought up mainly in groups begin with the opposite assumption; what the adults are saying is likely not to be personally relevant to them and can be ignored.

In the beginning, for a child to learn in a group setting, he has to feel engaged in a relationship with his teacher. He has to feel he matters, just as he does at home with his parents. First-year teachers describe their dilemma when faced with ten children, at least, who want to tell their story, or give their answer. It is a real juggling act to make everybody feel heard without crushing those who do not get a turn, and also noticing those who do not put themselves forward. For the child, too, having your hand up and not being asked can feel like not being seen. It is as if, after being their parents' precious child, they suddenly become invisible. The child may face strong inner conflicts, a simultaneous pull forwards and backwards in time.

DON'T TREAT ME LIKE A BABY

In the early years and at the beginning of school, children are preoccupied with rules. There is both a need and a reluctance to follow them. Children will constantly ask, 'Am I allowed to . . .', often to the irritation of their parents, who feel they ought to know by now – for instance, asking, 'Can I go to the toilet?' when they are home. It is as if they ask because they need to feel they have permission, they are observing a rule. It is a request for structure. It is also a way of managing the two different circles of home and school, teasing out what they can do where. Many a mother gets called by the teacher's name and vice versa.

During the school day, children have to listen to their teacher, to conform, to observe more rules than there are at home, and to fit in with the group. They may be successful at this and come home feeling very independent and rightly proud of their achievement. If this is not recognized at home and they are treated just as before, they will feel as if their 'growing up', still very precarious, is diminished. So they will often rebel. 'Don't treat me like a baby!' is a frequently heard cry throughout the primary years. They wish to leave babyhood behind, become part of the bigger picture. In their eyes, being a baby often means being silly, messy and out of control. Adults will reinforce this view by criticizing them with phrases such as, 'Don't be such a big baby!' At times, it is hard for the child not to feel like a baby, and being treated like one may threaten his efforts to function in a more independent way. This is when the parents have to hold back, say no to themselves, to their habit of seeing the child as static, still the baby at home, and recognize his forward movement.

The child who has not found it easy at school and has struggled throughout the day may want to rebel against the rules at home, where he feels it is safe to do so. He may have been very 'good' at school but come home and shout, kick and have a toddler-like tantrum. Some children will be glad to be babied when they are home – they regress and ask parents to do all sorts of small jobs for them, such as cutting up their food; this may lead the parents to worry that their child will never grow up.

IS THIS THE SAME CHILD?

Most children behave very differently at home and at school. Often the parent cannot imagine that the child they know is the same one other people talk about.

Peter, who is five years old and has just started school, always came back in an excited and irritable state. On the way home, he demanded a drink and something to eat, apparently unable to wait the ten minutes it took to get home. Crying, nagging and tugging at his mother's arms, he whinged and whined. His mother, who had been looking forward to seeing him at the end of the day, would quickly get fed up and worry about his behaviour. Why is he acting so spoilt? If he cannot wait this short time now, how on earth does he manage at school all day? Is he this difficult with the teachers? As a first strategy she brought a drink and snack with her when she collected him. This seemed to appease him for a while, but then he started to be upset by other things – impatient if she did not understand what he said straightaway or if she did not appear sympathetic when he told her of an incident at school. He was very grumpy and at the end of his tether and she felt everything she did was wrong or at least never quite right. At home, things did not get any better: Peter had tantrums, threw himself on the sofa if denied a request, hit out, kicked and even occasionally tried to bite his mum. Very worried, she talked to the teacher.

She learned that Peter was fine at school, very studious and well behaved, he listened and concentrated well, made friends in the playground. He got tired easily but other than that the teacher had no concerns about him at all.

In discussing this with me, it was apparent that his mother could hardly believe that it was the same Peter at home and at school. In fact, the Peter at school sounded more like the Peter she knew. There was a tremendous sense of loss of *her* Peter, the little boy she had spent so much time with, enjoying playing, going to the park, painting, visiting friends. She would even ask him, 'Where has my Peter gone?', a common question among parents who feel their children are changing.

As a way of understanding what was happening between them, we tried to imagine what it must be like for Peter. Just as she missed him and their closeness, Peter, too, must be feeling quite bereft. The mummy who was always there when he hurt himself, when somebody upset him or he had a question to ask, was not with him at school. However, he did also enjoy school and meeting friends and having more contact with a new peer group. It seemed as if he could hold himself together during the school day, partly through enjoyment and partly through wanting to be good and do well. He also had a positive experience at home, which gave him an optimistic approach to his teachers, so that, on the whole, he was happy at school. But by the end of the day, he was exhausted and more vulnerable. Also, seeing his mother put him back in touch with missing her and feeling the strain of the day, and to some extent blaming her for it. How could she put him through this? So on the moment of contact all these feelings came to the fore and got released in a flood of emotion. At this point, it is a bit as if the independent part of the five-year-old managed school but the more dependent, infantile (hungry baby) part has come home. This split often remains well into primary school and in fact recurs throughout life at particularly stressful or transitional times.

What could be done to help both Peter and his mother? The first point was to check Peter's state at school. The concern about a child reflecting badly on us can interfere with our sensitivity to their plight. So we established that Peter was really quite happy and settled at school. His mother's initial worry about how he might be behaving there abated. Although a relief, the fact that he was happy added to his mother's feelings of loss, and made her feel a bit jealous of the teacher, whom Peter praised and seemed to like very much. His mother's image of herself as the good provider was undermined; she felt replaced and this made it harder for her to cope. So we talked about what helps children to get on well at school and to make a good contact with their teachers. It was clear from Peter's history and how his mother interacted with him that he had a very secure base at home. I pointed out that it was this which gave him the confidence to go forward into school, believing people would be there to assist him. Her good relationship with him is part of what helped him have a good one with the teacher. The knowledge that she had provided him with the means to cope without her made his mother feel better about herself. It re-framed the picture from one where she felt rejected to one where she saw herself as helping him move ahead. Once we were able to look at their shared feelings about having lost each other, she was again more in touch with a view of herself as somebody good who is being missed. This helped her to be more available to Peter's feelings of hurt and disappointment. Daily life had changed for both of them and we were able to see that there had to be a shift and a development in the relationship. Yes, they had lost something, but they also had the opportunity of gaining something new.

With the untangling of feelings and making some sense of what might be going on, things improved. Peter still came out of school crotchety, but his mother was prepared for it and dealt with his tiredness and hunger by being a mummy who was there for him, cross feelings and all. She was able to feel less criticized and hopeless and he gained strength from her resilience and patience. With time he got less tired and was able to pace himself better at school, perhaps relaxing a bit and not needing to make such a huge effort to keep himself together there.

From their story, we can see how worries, separations, feelings of rejection, can make us too hurt to recognize the other person's experience. The two people become involved in a cycle of dissatisfaction and criticism of each other, and paradoxically grow more distant while mourning the loss of their closeness.

One of the purposes and effects of saying no, of setting limits, is to promote growth and development. We must therefore be able to adapt to movement when it occurs. However, it may not always take a form we expect or are comfortable with, and then our task is to hold back, to say no to ourselves and to our urge to force our approach or visions on our child. We must allow the child his own method of integrating what has come his way.

Accommodating what appear to be contradictory aspects of the same child can be very strenuous. It is hard to reconcile the six-year-old who acts so composed at times with the demanding baby he sometimes becomes.

Clare, just turned five, was on holiday in Italy with her parents and older brother and sister. They visited museums, ate at restaurants and were generally out and about. Her parents were impressed by how well she participated. The only problem was that she drove them crazy with constant requests to look after the new Mickey Mouse toy she had been given. They would have to find a seat for Mickey in the café, make sure he was contented, explain what was being said in the foreign language, and so on. Panic would break out if Mickey's blanket was lost. The whole family began to get furious at Clare's requests on Mickey's behalf, wanting to shout, 'For Heaven's sake, it's just a toy!' It was only when they saw her preoccupation with Mickey's comfort as a sign that she herself was feeling at sea – away from home, her bed, her routine – that they felt less irritated. They were able to feel sympathy for Clare's *dépaysement* which she had not expressed directly. Once they felt it was Clare they were looking after when thinking about what Mickey might need, life became easier.

During the primary-school years children will fluctuate from autonomy to reliance, from being reasonable to having a tantrum, from being confident to being very insecure. As a parent, it is hard to get the balance right between supporting their independence while staying in touch with their neediness without making them feel infantilized. When and how we phrase the noes will be important. We can make the child feel belittled and humiliated, or make him feel protected and safe. As my thesis in this book is that noes help to build and not destroy, it is important to think about the child's perspective, to consider the impact of a 'no' on the child's growing self-confidence.

REASON AND LOGIC

The French say that at seven children reach *l'âge de raison*, the age of reason, that they are then able to act rationally and consciously.

You do see children of this age discussing and arguing for their point of view. They pride themselves on being logical and hate to be thought of as unreasonable or ruled by emotions. It is important for the strength of their emerging independent self that this developmental urge be acknowledged. Yet there is a split between their reasonable side and their more passionate, infantile side. My daughter Holly, aged five, got into a terrible argument with her good friend Sassy about the origins of the world. Sassy insisted God created the Universe and Holly was just as adamant that it was all down to the Big Bang. They got quite upset with each other, as if what was in question was their position as 'someone who knows' rather than the actual answer to the origins of man.

You will often hear very young children debating major issues which they may have little grasp of, but which matter to them as important aspects of being part of the world of thought and language. Another wish is to be taken seriously and to feel confident that you are right. The child may bring evidence which has a logic of its own. Joanne, a school-friend aged six, was sure that there were no elephants in the UK. When the other children said they had seen some at the zoo, she refused to believe them. When asked to referee, I tried gently to say that elephants did not come from the UK but that there were some in the zoo. In that way Joanne *and* the others were right, depending on how they looked at it. This, however, was not enough for Joanne, who used as her ultimate weapon the fact that she knew she was right because her mother's hairdresser had said so. Puzzled, I asked how come the hairdresser was an authority on elephants. She answered: 'She knows more than you do because she's older than you!' All children are told at some time that elders know better, so one can see how she used her own logic to argue her case.

The issues of being right, of fairness and justice, take on great importance. As the child rejects baby aspects of his personality – the emotional and unruly parts – he sets great store by his ability to reason and discuss. A disagreement can therefore feel like an attack on his integrity. He may become absolutely distraught if he feels misunderstood, as if the very fabric of his being is in question. It is a time when children are learning to get on with others and developing a conscience. On the whole they desperately want to be 'good' and can be very harsh on the aspects of themselves and others which they consider 'bad'. Bruno Bettelheim writes about how we unconsciously reinforce this split:

> There is a widespread refusal to let children know that the source of much that goes wrong in life is due to our very own natures – the propensity of all men for acting aggressively, asocially, selfishly, out of anger and anxiety. Instead, we want our children to believe that, inherently, all men are good. But children know that *they* are not always good; and often, even when they are, they would prefer not to be. This contradicts what they are told by their parents, and therefore makes the child a monster in his own eyes.

Arguments at this age, therefore, carry a great emotional weight; they are often not just about the issue at hand but seem to put the whole identity of the child into question. This needs to be remembered when entering into the debate. When you contradict him, does he experience it as a specific moment or as a judgement on who he is? Similarly, when he behaves in a fashion that upsets you, is it a one-off or does it stir up anxiety in you about what kind of person he is?

Conflicts

Conflicts frequently arise around mealtimes, bedtimes, how much TV to watch, getting homework done, getting ready in the morning, comparisons with siblings and friends. Let us look at some examples, examine what might be happening for the parent and the child, and think of ways to understand and deal with the difficulty. All these interactions are to do with relationships: each party brings his or her own contribution, which has an impact on the other. Inevitably, therefore, at times we will need to say no to the child, at others to ourselves.

SIBLING RIVALRY

With the start of school, the child is no longer mainly at home, in a world principally defined by the family, but has to integrate into a wider group. How does he find his place? Does he push his way to the top, being naughty in a way that excites and attracts others? Does he bully in order to feel less scared himself? Is he the little one at home, and so feels he has to fight for his space at school? Does he take a back seat and watch others rather than actively joining in? Does he find it hard to keep up and get a bit left behind? There will be as many reactions as there are children. As parents, we have an important role in helping them negotiate this transition.

Sometimes the tangles at school are a reflection of what is happening at home. At school children repeat battles they are having with their siblings. Alternatively, home may be the ground on which school conflicts are practised and worked through. The child's relationship with his brothers and sisters can become very intense, and rivalry more pronounced. In Chapter 2 we saw that toddlers have a struggle getting used to a new baby, or finding their place in a family with older siblings. This endeavour carries on with every age group. At primary school the issue of one's place in the group stirs up feelings of insecurity, and this can spread to the home. There may be many cries of favouritism, complaints that a parent prefers the older or younger sibling. When faced with these accusations, it helps to remember that the child is having to establish himself elsewhere and may be looking to you for confirmation that he has a secure place with you.

George, aged eight, is the middle child of three. His mother spends a lot of time with his three-year-old sister, whereas his older brother, who is thirteen, goes out with his friends at weekends. George feels very left out and gets into big battles with his mother. He tells her it is unfair that she won't play with him yet she won't let him go out either. He feels dissatisfied and often hits out at his little sister or pesters his older brother, not leaving him alone.

We can see that George feels his siblings have special privileges and he has none. It is a difficult time for him: he is too old to need to spend as much time with his mother as his sister does and yet too young to be allowed out with friends on their own for extended periods of time. His mother has to find the appropriate limits and boundaries for him. He may need to have some of the privileges of his little sister as well as some of those of his elder brother. He will also face restrictions specific to him.

Many parents and children judge fairness on whether everybody has the same. However, it is clear that this can never be possible or appropriate. The parent has the much more difficult task of finding what is right for each child. Negotiating everybody's wishes and needs, including our own, and sometimes leaving everyone dissatisfied, is part of everyday life.

It is helpful for the child to feel free to express his complaints and have them heard. This does not mean he will automatically get what he is asking for. Moreover, it is not helpful for him to hurt or consistently undermine his siblings. The mother who stops her child from being a bully, or from taking over every situation, is helping him keep his angry feelings within bounds. The child who is allowed to be destructive will feel unsafe, as we saw in Chapter 2. He will feel miserable about being awful as well as frightened that no one will protect him if the need arises.

This is also true of more covert ways of dominating situations. An older child may over-baby his younger sister, treating her like a little doll. This patronizing attitude may be a way of managing his own aggression. It may also be an attempt to distance himself from his own more infantile feelings by overdoing and caricaturing the grown-up role. A certain amount of this type of behaviour is bound to happen. However, if it dominates the younger child's life, if the mother is unable to intervene, it will do neither child any good. It stifles the little one from finding her own wings, and permits the older one to take over, to be a tyrant.

Another common dynamic is when one child is obsessed by what everyone else has got.

Greg, aged nine, was referred to me because he was very disruptive at school, constantly getting into trouble for picking on other children and interfering with their work. In his sessions with me, he was so curious about who else came to see me that he could hardly think of anything else. He would examine every inch of the room for traces of other children. He set up elaborate traps, with strings and Plasticine, to check whether anyone came between his sessions and catch them out. He would badger me to get information about my other patients. His obsession led to very repetitive play and an inability to shift. He spent all his time captivated by what he imagined everybody else did or had. This left little room to think about any other aspects of his world. For a long while he brought in flower heads, full of seeds, which he opened and examined. It was as if he felt me to be like that, full of seeds, other babies, which filled me up and left no space for him. By taking up the entire session with his pre-occupation with others, in effect he did fill up our joint space with others. It was hard to make him feel unique. When he asked what other children did here, it took a long time to be able to think with him about what *he* might like to do.

Such children, who do not feel secure that they have a solid place in your mind, who feel you are always preoccupied with someone else, are then driven to compete with or get rid of other children. This leads to problems at school, as with Greg, who actually attacked the other children. A child like Greg may hear a 'no' as a rejection of him in favour of someone else. It was important for him that I preserved my care and interest in him, and expressed it clearly. I had to spell out that, 'No, I will not tell you what others do here, but I am interested in what you think they might do or even more in what you would like to do.' It was also crucial that I showed understanding of how awful he felt about not knowing, without breaching my boundaries about the other children. By not talking about them to him, not revealing who they were or what their concerns were, I was giving him the idea that his space with me was also protected. I explained that I would not allow him to break into their space, even under immense pressure, but this also meant that none of them could break into his. This eventually gave him some taste of a time just for him.

In a case like Greg's, it is not curiosity which is driving his questions – a quest for knowledge which one would want to encourage. Rather it is an intrusive, nosy ferreting about for information he could use against others. These kinds of questions arise in families, too, and we need to learn to preserve some privacy for each member of the family. Not everybody has to be involved in every conversation. Learning to tolerate this is a great achievement, and we will see in Chapter 4 how hard it is in adolescence.

Making a stand in the family for treating each other with respect, not being consistently unpleasant or rough, gives the child the means of coping at school with difficult feelings and behaviour. Saying, 'No, you cannot behave towards me (or your sister or brother) in this way', spurs them on to finding other means of dealing with frustration or upset. It increases their emotional vocabulary and should help them find appropriate strategies when faced with provocation in their peer group. In Chapter 2 we saw how making a point of keeping your own self-respect, not allowing yourself as a parent to be abused, sets an example. It shows that you believe you are good enough not to be treated badly. This helps the child do the same. If he gets teased at school, left out or picked on, he will have a model of how not to be the victim. We have all seen how some people, children and adults alike, seem to invite derision, always casting themselves in the role of the victim. During these early years children can form a strong sense that they deserve better and will not be put upon. This is not done by becoming the strongest one yourself, like the bully, but from believing you are worthwhile.

A SENSE OF TIME

Once they start primary school, time takes on more meaning for children. For babies and under-fives, it is inseparably linked to the child's experience. A hungry baby waiting for a feed may feel five minutes is an eternity. A three-year-old enjoying an afternoon with a friend may have a tantrum when it is time to go home because he feels he has only just arrived. At school, however, the day is organized into recognizable chunks of time – lessons, breaks, lunch. This lesson may feel much longer than the last, but the child knows that, in fact, it was the same length. There is also a growing concept of keeping to a timetable. Getting to school, having your supper, going to bed and many other activities fall within a specific framework.

As parents, we start to expect and demand that our child fits in with requests within a limited time span. However, children, particularly at this age, seem to have a selective view of time. When it is something they want, the key word is *now*; when asked to do something, it will be *later*, which often means never. This is another reflection of the ambiguous position the child is in – not a baby and not yet too self-reliant. He is learning more self-control but waiting is still hard. Wishes have a pressing immediacy. In saying 'wait' at this stage – saying no to an instant response – we are asking them to hold on to their want, or find ways to satisfy it themselves. It is important that they should not wait so long that the want disappears, or we would be dampening their zest for life. But they do need to learn to manage a space in between. Often the waiting gives rise to irritation, loss, anger, hopelessness. These difficult feelings are all part of the human emotional repertory and being familiar with them is no bad thing. It goes towards making us more complete.

Waiting is an interesting experience to look at, full of ambivalence. You wait because you wish for something. In the wait, however, this thing you wish for may turn bad. For example, recently my daughter Holly wanted a cuddle before going to bed. I was in the middle of a task and could not go straightaway. Twenty minutes later she came to me, furious, and left a note which said, 'It is 9 p.m., I hate you!' In the time that had elapsed I had turned from the nice mummy she wanted to cuddle into a nasty one she hated. We recovered easily, and the cuddle was enjoyed by both of us.

Repeated simple experiences like this make for the necessary distance between parent and child which allows them separate lives. From a more philosophical point of view, though, these moments also teach us about ourselves, how we deal with absence, waiting, gaps. Do we hold on to the good missed person or do we turn her into someone bad? How do we represent in our minds somebody who causes us pain? Can we recover when she returns? These questions are pertinent during our entire lives and I will return to them in chapter 5 about couples. In the case of my daughter, Holly was in touch with her ambivalence. Ron Britton, a psychoanalyst, writes: 'When we acknowledge we hate what we feel to be the same person as someone we love, we feel ourselves to be truthful and our relationships substantial.'

Looked at this way, saying no is not cruel, it is a necessary part of being separate. If you are not separate, you cannot have a relationship. Parents who fit in with their child's every wish are giving him the illusion that they are extensions of him, that only his needs matter. As we get more involved with each other, we learn to cope with our differences as people – what I like may not be the same for you. As the child's capacity to tolerate this grows, so does his awareness of others and their feelings. These will be important skills at school and in life. Many adults have not mastered the idea that you do not deal with differences by trying to make the other person like you, and this is visible in many marital problems! The foundations for the move away from selfishness are laid in childhood.

Given that time is experienced differently by all of us, how do we accommodate these variations? Your child has asked you to play with him and you have been sitting together at the Nintendo for half an hour; you then decide you want to do something else.

He gets upset with you, saying you never play with him. You reply that you just have. He retorts, That doesn't count, we only just began. Both points of view are true emotionally. You ask your child to pick up his clothes and put them away, he says he will do it later. By the end of the evening, they are still lying on the floor. He intends to do it, but his later is not the same as yours – it may be in a few days' time as far as you can see! You have to negotiate.

All these mismatches, which are part of daily life, help us to learn to live with others. We are dealing here not just with behaviour but with emotional understanding, the realization that their experience and ours may not correspond.

EXPECTATIONS

We tend to try to understand others by putting ourselves in their place. This is a helpful way of imagining what they might be feeling, wanting, communicating. However, sometimes we are very different from others; our own children may make us wonder, 'How can they be so different from me?' Then it is important to look at them and see what they are telling us, rather than what we would like to hear or assume they mean.

Mrs W always worried that her daughter Zoe, aged ten, would become a couch potato. Mrs W watched TV as a way of totally relaxing, a comforting, mindless activity at the end of a hard day, when she would happily sit and watch anything for hours. She was therefore very reluctant to let her daughter watch TV, as she was sure Zoe would stay glued to the screen and do nothing else. Zoe was allowed no TV during the week, which gave rise to many battles. Mrs W stuck to her position absolutely. Mrs W's sister, who had a different attitude and with whom Zoe spent a lot of time, especially in the holidays, told her that when Zoe visited them, she loved to unwind with a half-hour television programme, after which she was very keen on playing with her cousins for the rest of the evening. She convinced Mrs W to try letting Zoe watch a bit at home, which she did. As predicted by her aunt, Zoe did like a limited amount of TV but was quite happy to engage in other activities afterwards.

It was Mrs W's assumption that Zoe would be just like her that informed her stance. Although it is important to be firm when saying no, it is also crucial to spot the occasions when a no may not be helpful or necessary. Here it seems Mrs W was trying to deal with her inability to say no to herself, when she watched what she considered to be too much TV, by making a stand with Zoe. Zoe in fact was able to limit herself.

Billy, too, is very different from his parents who are both professionals, very organized and efficient.

Billy, aged seven, is always late in the mornings, hunting for the shoe he has misplaced or having to dash back into the house to pick up some item for school he has forgotten. Billy's parents find it hard to accept that he deals with life very differently from them. They get irritated with him but do not take steps to solve the problem. So, in the evenings, when Billy wants to play till bedtime and is then too tired to tidy up or prepare his school things for the morning, they let him off.

Billy's parents are not in touch with his different pace and assume the morning will run smoothly for him, as it would for them. Even though the same thing happens day after day, every morning is experienced as a surprise; they think, 'I can't believe it, he's done it again!' or, 'How can any child of mine be so disorganized?' It is as if they cannot really accept that Billy is not like them and cannot organize himself easily. He takes more time, needs to be checked and has to do things in advance.

We often find ourselves in similar circumstances, being surprised by events which occur regularly. We then need to think about why we do not learn from experience and why we repeat patterns which are unhelpful. Here the parents' reluctance to shake off their assumption that Billy is the same as them leads to an inability to say no to him in the evenings, when he wishes to do other things rather than getting himself ready for the morning. They would have to recognize his particular problem in pacing himself and ensure he prepared his school things in advance, even if this meant a confrontation in the evening. Those rows might be preferable to the daily frustration and chaos they all end up with.

More often our expectations of our children determine where we make a stand, when we feel limits are necessary. We tend to enter into conflicts when our expectations contrast with our child's behaviour or attitude. We want to say no to their way, and impose ours. Who does the child aspire to be like, who do we urge him to be? What standards do we use to assess success? Particularly with adolescents, we invest a lot in our children's attainments, yet, even at primary school, when children start to learn, grades may be given, tasks or homework are assigned and deadlines are to be met, the spectres of failure and success raise their heads. Are we looking for perfection or 'good enough'? Do we want the child to excel in relation to others or do his best in relation to himself?

After school it is common for a child to be helped with a maths problem which he simply cannot grasp by an exasperated mother who does not understand why something so simple is not clear to him. She gets angry, as if the child is refusing to learn, is being stubborn or provocative. He gets upset and stops listening. Frustration rules and a cycle of misery is entered. This can be a regular scene and one which can trap both parent and child. The mother thinks, 'If only he would listen to me', the child feels she is disappointed and that he is stupid. He may protest and want the mother to leave him alone.

Our expectations can interfere with the pace and manner in which our children best learn. We feel inadequate when they do not understand or make progress. We worry about how they will do in the future. We feel rejected when they do not want our advice. The focus moves away from the issue itself, such as the maths homework, to the relationship between the mother and her child. A friend of mine is teased at home for stating in despair, 'People pay thousands of pounds for my advice. You can have it for nothing and you never want it!' It is at times like those when the mother needs to make a space between her and her child, to say no to being drawn into a conflict by her frustration. She is in a better position than the child to keep to the task in hand and not let her own mind drift to questioning herself as a mother and him as a son. If she can achieve this, she will be more able to be helpful and to stop herself rushing him.

In certain cases the child may be worried about his ability and what is asked of him, which may be more than he is able to deliver. We can see how this affects his capacity to take things in. Here is a short extract of a school scene, involving Alex, aged eleven, observed by his support teacher:

> The children come into the classroom and sit quietly. Alex is playing with some coins. He is moving around and not listening. The class teacher asks him to sit down. He sits properly and listens for a short while as she explains their task. He asks if they have to write things down. The teacher gets irritated with his query: 'Alex, we have been doing this for three months now and still you don't know! I don't have time to answer silly questions!' Alex starts flapping his hands, playing with his coins and passing them from one hand to the other. He blinks and turns his head away when I try to talk to him.

Alex is anxious and unsettled. He fidgets, wants to participate but does so inappropriately, by asking a question he should know the answer to, thus irritating his teacher. He finds the undertaking daunting and is easily distracted. His worry about not being able to do the work leads to a lack of concentration, he does not listen and therefore starts the task with a handicap which is then hard to remedy. His fear of failure stops him from learning.

The next example shows what happens to children who really cannot keep up with what is expected of them. We can see the defences they resort to and the nature of their fantasies. Lee, aged nine, is in a mainstream school but also receives help from a support teacher. He has great difficulty in the basic skills of writing and simple arithmetic. He is constantly being told off:

Lee enters the classroom in an excited mood, greets everyone loudly and jokingly pushes and kicks other children. They try to ignore him. The teacher explains the task for this lesson: they must write a letter to the owners of an imaginary place called Fantasy Land, which they wish to visit. They have to say why they should be invited. Lee laughs and points at other children. The teacher asks him to get on with it, and his support teacher, Tanya, goes to sit with him. There is not much room for her. He tells her rather provocatively, 'Sorry, there is no place for you to sit!' She feels irritated with him and says, 'Well, sorry for you!' He answers, 'What did you say, Tanya? You feel sorry for me?' She explains that if there is no place for her, she cannot help him. He then aggressively pushes his chair to make room. He hits Khalid, the child next to him. Khalid asks him, 'What are you doing?' Lee answers, 'It wasn't me, it was Tanya.' Tanya tells him it is time to concentrate. He swings on his chair, and whispers to his neighbour. They both giggle. Tanya tells them it's time to work. She asks Lee what he would tell the owners of Fantasy Land. He replies, 'I don't know,' and swings more vigorously. The others are now writing and he looks at them anxiously. Tanya asks again, 'Shall we try and think what you could say about Fantasy Land?' He answers that there is no such place. Tanya says, 'Let's pretend and use our imagination.' 'But the place does not exist,' he reiterates. Tanya says they can try to invent it. Lee looks at what Khalid has written, and whispers something to him. Khalid gets very upset, reporting to Tanya that Lee said he saw Khalid's mother coming out of the Oxfam shop with a bag full of clothes. Khalid protests, 'It's not true!' Lee laughs and says, 'At least I don't buy my clothes at Oxfam.' They argue and Tanya tells Lee off for disturbing Khalid and upsetting him. Lee denies it. Tanya tries to get him back on task and begins writing to get him started. He is amazed at how quickly she finishes a paragraph. He tells another child, 'Look, Tanya is writing out of blood, you know that the best way to write is to cut your wrist and as the blood flows you write pages and pages.'

In this extract we can see not just what a pain Lee can be, but also what a hard time he has at school, how painful it is to be him. He enters full of bravado but is quickly crushed when Tanya comments, 'Sorry for you.' We also see how his attitude provoked a cross, tit-for-tat response from Tanya, which she later modified into something more thoughtful. This points to how difficult it is to stay aware and considerate with children who constantly push the limits. He took the remark as a put-down, felt humiliated and proceeded quickly to hit another child. Many children like Lee seem to lash out randomly, for no apparent reason. Here Khalid was puzzled by Lee's behaviour and asked him what he was doing. He could not see why Lee had hit him, totally without provocation. Yet Lee's behaviour did have a meaning: it was related to his anger with Tanya, not Khalid. He was upset and hit out at the first available target. He felt hurt and wanted to hurt somebody back. He also blames Tanya, as much younger children do when they hurt themselves and say, 'Look what you made me do!'

When faced with his work, Lee doesn't know where to begin, engages in delaying tactics, trying to change the subject and distract other children. Once they are all at work, he feels very isolated. When Tanya asks him to imagine the Fantasy Land, he is absolutely unable to do it. He cannot deal with things that are not visible and concrete, he cannot let his mind explore and create. This is like the babies we looked at earlier who are always active and have never let their mind wander. He is at a loss and anxious. When he looks at what Khalid has written, he feels inadequate and envious. So he begins to torment him. It is as if he cannot bear to be upset himself and must make someone else feel bad. He picks on Khalid to stop him from getting on with his work. His story about Khalid's mother is not just aimed at hurting Khalid, but can also be seen as the expression of the cruel, mocking voice he carries inside himself. He cannot work by himself, first-hand as it were, and has to pick up bags of second-hand thoughts from other children and teachers.

Experiencing Lee's cruel taunting of others, we can imagine what this same attitude would have to say about him. He is a child who feels stupid and wants to make others feel as bad as he does. The harsh, stringent expectations with a mockery of his inadequacies come from inside him, as well as outside him, occasionally, from the teachers. He tries to protect himself from feeling incompetent by not involving himself with his support teacher or any one who reminds him that he does not know or cannot do a task. Unfortunately this also stops him from opening himself up to help. When he sees how easily Tanya writes, his fantasies become morbid, about cutting and blood. We get the picture that writing for him is a nightmare, that you have to suffer, maybe even die, in order to do it.

For such children, being told 'no' is crushing; it echoes the nasty critical voice they carry around. The first step is to build up their self-esteem before they can become accessible to learning. Lee is like an older version of Paul, whom I described in Chapter 2. He cannot bear to be in touch with being vulnerable, he copes by hurting others and cutting himself off. His overt stance is brash and confident. This invites people to react harshly, to be angry and set rigid limits. They reflect his inner world, the contemptuous spoiler within him. The challenge for adults with children like Lee is how to set limits which feel safe rather than punitive. He needs to begin to feel that although he cannot manage some things now, there is hope that he will one day – that being unable to do his writing does not mean he is a total failure, to be humiliated and ridiculed. It is also important to let him know that although you understand how he feels, you will not let him hurt others. This will give him hope that you might also take a stand against the part of himself which mocks his own failings. A child like Lee will not just grow out of it and does need special help emotionally as well as educationally.

The following example from a similar setting shows how boosting confidence can facilitate learning. The way the limits are set do not belittle the child. Prithi is ten years old and has extra help at school. She, too, has great problems in reading, writing and simple school work. She tends to want the teacher to do all the work for her. Her support teacher, Liz, writes:

I tell Prithi that we have to read a story about dragons and then answer some questions about it. I read the story but she does not concentrate, she is distracted, plays with her pencil, moves about in her chair and tries to distract me. I say that she seems worried about all the facts we have to know about dragons in order to answer the questions. Prithi replies: 'Liz, I think the problem is comprehension, I mean, you know, the other children can do this quite fast if they don't have learning difficulties. If you are intelligent, I mean, no problem, you see?' I say that I think Prithi is worried about not feeling as intelligent as the others. I add, 'I think your difficulty is nothing to do with intelligence, you are an intelligent girl. It is not that you don't understand what I am saying about dragons. If I read to you that dragons in China are kind and friendly and bring good luck, whereas in England they are ugly and are meant to guard treasures, can you understand that?' Prithi nods. I say, 'See! It is not that you don't understand. I notice that often it is difficult for you to concentrate on what we are doing.' Prithi calms down and asks me to continue. After a little while she asks if she can read the story about the dragons to me.

In this extract we can see that Prithi feels she is not up to the task and compares herself unfavourably to her peers. We know from other reports that she tends to give up and usually asks Liz to complete the work. However here, when Liz talks about how Prithi feels and gives her hope that she might be able to understand and participate, Prithi is able to have a go by herself.

I have given three examples of children with learning difficulties to highlight in an extreme form what happens when expectations do not match what the children can achieve. The examples also show the different ways in which they deal with these situations. Children with less severe difficulties will show similar adaptations, and we should consider whether what we ask of them matches what they can produce. In our dealings with them, we need to be specific, not global, so that they feel helped with a particular difficulty rather than judged. We need to be aware of the insecurities faced by children when they do not have the answers. Even as adults, it is not easy to carry on trying to understand something we find hard, or to remain in a position of not knowing.

Many conflicts and arguments are based on mismatched expectations. We may get angry with children because we feel they are not trying hard enough. The actual problem at hand – the task that has been set – disappears from view and what takes over is a difficulty in the relationship, whether this be with a parent or a teacher. If we are able to stand firm, say no to the child who tries to distract us and keep him on track, we are more likely to stay focused. We will also be clearer about what is interfering with his learning, as Liz was with Prithi, and so we may be able to keep the child with us, not lose him to his own anxieties. Many children drift away easily; they need a firm hand to keep them present. We may also need to say no to our own ambition, and slow down to join the child at his pace.

On the whole, as parents, we are out of touch with what is expected from primary-age children, unless we deal with them professionally. What was expected of us when we were at school is not the same as what they do today. We are often surprised that a teacher's feedback differs from our assessment of our child's ability. In the later primary years, particularly, we may compare what the children are doing to how we would do it now. So a project may seem very flimsy, or a science experiment unclearly presented. Many parents have to resist the urge to take over, offer a plan of how to do the work, or do the research for their child. In our wish to be the 'good' parent who helps with homework, we are in fact interfering with the children's pace and capacity to learn for themselves. They have to do it by themselves, if we want the skills to last. In these circumstances, we must learn to say no to our wish to be involved. I am not saying children should be left to struggle miserably by themselves, but we should be responsive to their need for us, rather than to our need to feel good or in charge.

Some children do not want to work, they want you to do it all. They may be like Prithi, and feel they cannot do it, or they may just want to be babied. They may feel it is a way of keeping you with them, if they act helpless and in need. They may use many distractions to keep you from saying no to their wish to be treated as babies all the time. You will need to work out how to accept their expression of infantile need and do what you can to make them feel better, while still expecting them to stretch themselves at other times. It may be that they have done enough being 'grown up' at school and just want to flop and regress at home. However, if they want to stay babies all the time, you will not be doing them a favour if you join in. You will be agreeing that being the age they are now is not as nice as being younger.

In a similar vein, we sometimes see children who are treated not so much as babies but as terribly special, as little princes and princesses. Frequently the expectation from their parents and themselves is that everyone else should do the same.

Carla, eight years old, is an only child. She is petite, rather pretty and looks younger than her age. She is a bright child who does many extra-curricular activities. She goes to ballet, tennis, plays the violin. Her mother is often at her school, going on outings, helping whenever there is a request for parents to join in. Carla speaks like a mini adult, criticizing other children for being messy or a teacher for being late. She is always keen to tell everybody how well she has done in any test and makes a point of being well behaved. If she gets into an argument with another child, she is quick to go to the teacher and occasionally her mother, if she happens to be there. The mother, in turn, will intervene on her behalf and scold the other child. Carla is very charming and polite to teachers and gets on well with them but is deeply unpopular with the children.

It is obvious from this description that this state of affairs is not helpful to Carla. At school it is not that noticeable to the staff, to whom she appears mature for her years, chatting to adults and seeming more composed than many of her fellow pupils. She has everything she could wish for, but experiences very few limits. She behaves in a way that really irritates her peers, being a goody-two-shoes, and always wanting to stand out. She has to feel central to what goes on, cannot truly join the group and take her part in it. Her expectation that she should be as special to everyone as she is to her parents will eventually prove to be a handicap. She is in fact being kept immature. By wishing to give her the best, her parents are depriving her of the essential experience of being ordinary, just one of the kids.

With certain children it is important to have high expectations of their capacity to learn and succeed. A pivotal piece of research divided children of similar ability into three random groups allocated to three different teachers. The teachers were told that one group of children were particularly bright, the middle group average and the third of below average ability. The research showed that the children's achievements matched the expectations. This is crucial information in dealing with children who are labelled as problematic for one reason or another. Certainly some children have more ability than others, who may require extra help. It is important to assess this carefully rather than work on assumptions or prejudice. Nowadays, for instance, an inordinate number of children from ethnic minorities, boys particularly, suffer from low teacher expectation. Similarly, the fact that a child lives in a home where a foreign language is spoken is frequently considered, in this country, as a drawback. It could just as well be seen as a great advantage. Many children in India, for instance, will easily handle two or three languages. High expectations, if these do not feel tyrannical, fill the child with hope and encouragement.

Some children may be very interested in learning and their expectations outstrip ours. This is exemplified by Roald Dahl's marvellous story of Matilda, the child who is so gifted and for whom books, learning and her teacher are a haven. Her parents are caricatures of greed and shallowness. The story can apply to most of us, too, in an indirect way. If children's interests differ from those of their families, they may not be valued. Even in a minimal way, a literary family may not follow up conversations and queries about science. They have to recognize that their child's talents lie elsewhere and resist overvaluing their skills as opposed to his. Occasionally a child has a true passion and talent for something completely out of the family's frame of reference; here the parents will have to struggle against their sense of unfamiliarity and uncertainty to allow the child to explore this uncharted territory by himself. Many children will also learn far more than their parents did at school, which is one of the reasons why education is highly valued as a way of ameliorating our circumstances. In such cases the family may not be able to help the child with the actual content of the work, but will play a crucial role in encouraging him with the process of learning.

GROWING INDEPENDENCE

We need to be flexible and change our expectations over time. Toddlers who rush into the house with muddy shoes and throw their clothes all over the floor will probably be scolded, but are generally treated with a great degree of tolerance – mothers pick up after them, even though they complain. A ten-year-old who behaves like that will not be very popular at home. The same expectations will apply to general tidiness, hygiene, and the belief that a child should be self-disciplined and self-motivated. A parent who constantly has to tell her child what to do or not to do feels like a nag. This is a horrible position to be in and the child will be expected as he grows older not to need that. The mother may have to resist the urge to keep on at him. She will have to give up being in charge so much.

A child who remains very reliant on his mother to provide everything is restricted in his range. He may want her to play with him or always suggest what he should do next. He may complain as soon as he is left to his own devices. He will regularly cry, 'I'm bored.' His imagination and creativity become stifled. In Chapter I, we saw how a parent's absence can lead babies to develop their inner resources. It is just as true at primary-school age: saying no to being there all the time can be very helpful for the child. Our job is not always to fill the gap, but we must also tolerate this uncomfortable position. Adam Phillips, a child psychotherapist, writes: 'Is it not indeed revealing, what the child's boredom evokes in the adults? Heard as a demand, sometimes as an accusation of failure or disappointment, it is rarely agreed to, simply acknowledged.' He adds: 'It is one of the most oppressive demands of adults that the child should be interested, rather than take time to find what interests him.'

A growing child wishes to have greater freedom, to venture out on his own, make his own decisions about which friends he would like to see. It is fundamental at this age to give him some freedom, for instance to go out on his bike, or to play with others on the street. He needs to feel there are not only restrictions. The importance of being with other children cannot be underestimated. The parent may have to go to some lengths to ensure that the child can see his friends. This may involve some extra to-ing and fro-ing and the parent and siblings may have to forgo some of what they would rather do. This realization should help the child accept the limits parents set him. Letting him go further and fend for himself a bit may be hard for parents; they have to trust him and the environment. In recent years this has proved difficult for parents. However, the alternative is to extend the period of dependency and restrict children's life skills, as well as to portray the world as a dangerous place. We need to weigh the benefits and risks carefully.

Parents have to balance a child's more infantile needs with his increasing demands for independence. This can cause friction. What do you do when your child begs for a pet, assuring you that they can be trusted to look after another creature, that they will be responsible, do what needs to be done even when they don't feel like it? Do you agree, even though you know that they might not be up to it? Do you enforce the rules of the agreement, particularly when the animal clearly is not being properly looked after? Do you take over, or do you allow the child to learn the cost of not fulfilling his commitment, even if this means the pet suffers? These are not trivial dilemmas. If we take over and do all the work, the child may never realize his contribution was not enough. He may forget what happened last time when he makes a similar request. If you do not look after the pet, apart from inflicting cruelty on the animal, you are also allowing your child to be destructive and at worst letting another living being die, with the guilt that follows. You may end up in the awkward position of having to push your child to fulfil the responsibility he took on against your advice. However, not allowing him to try might also be unfair; he may well manage the tasks and gain joy, a sense of responsibility and maturity from looking after his pet. The issue of growing trustworthiness and what are appropriate limits is not an easy one. The happy medium is probably to enter such areas with optimism but caution, aware that the child is aiming at things he may need help with and cannot master totally alone.

A similar question arises when children want to take up an activity such as a sport, drama, dance or an instrument.

Frank, aged nine, is passionate about swimming and good at it. He persuades his parents to enrol him at the local swimming club, which requires him to attend three times a week. He is keen to compete but resents the practice. His father has to nag him to go regularly, even when he doesn't feel like it. This causes a lot of arguments.

Having encouraged or responded to a wish of the child's, many parents feel it is their responsibility to see it through. My suggestion, at this point, is to consider what carrying on or stopping means to the child. Are you forcing them to do something they truly have grown to hate? Should you give in to a part of them that gives up too easily? I believe such decisions should be thought through carefully rather than taking the easy, quick exit. By spending time thinking, you are showing that taking things on and dropping them matters. This is useful in itself. It may be important to say no to the urge to give up when the going is a little hard, in order to develop a sense of sturdiness and optimism. If you give up easily, you may never know that you could have made it. Hopefulness is undermined.

In Frank's case his parents insisted he persevere until the end of the year. By that time it was clear that what he really enjoyed was the fun of swimming with friends, and his parents were satisfied that the club was not right for him.

BEING DIFFERENT

What we expect and allow our child to do will be partly determined by what seems appropriate for the age group generally, what we see other families doing. We then have to look at how this matches our own ways.

During the primary-school years, what other children are up to matters tremendously. You may notice that many children like collecting stamps, football cards, pictures of pop stars. They want to sing the same songs, watch the same programmes as their peers and discuss what happened in their favourite soap; they enjoy being part of a shared culture. Much of this is to do with feeling safe in the outside world. It is also about feeling they are no longer needy babies wanting their mummies but do very well with each other, thank you! Children observe and compare and most often want to be the same as each other. They lose some of their intensity and take comfort in structures and stereotypes. Their activities often appear boring to adults or teenagers.

Parents can run into difficulties when they do not fit in with the norm. How do you say no when Jenny tells you everybody else in the class can go to Alice's sleepover, or that all her nine-year-old friends are allowed to walk down to the shops by themselves? As always, you need to think and examine your stand and why you are making it. Is your instinct to say no due to your fear of letting Jenny go? Are you worried because you do not know Alice's parents? Does your cultural background clash with Alice's family's? Do you know that Jenny is crotchety when she is tired and worry she'll cause problems at Alice's home? Only once you have carefully weighed all the pros and cons can you make an informed decision.

Mrs G has four children. Her youngest, Jane, aged eight, has been invited to stay the weekend with friends in the country. Mrs G is reluctant to let her go, worrying that Jane will feel anxious at night. She also does not relish the thought of driving a couple of hours to collect her if she gets upset. However, Jane is so keen, her mother decides to let her go. In fact, Jane has a wonderful time and is happy to repeat the experience. Mrs G is surprised because she has always thought of Jane as the little one, who needs the comfort of the family home. Usually bedtime is an issue and Mrs G thought Jane would find it hard to settle. She is relieved Jane had a good time but also a little sad at the loss of her baby, at the realization that a phase is over.

We can see here that Mrs G was prepared to say no to her own wish to keep Jane tied to her apron strings, to allow her to venture forth. This is a mixed experience for her: she has gained a more self-sufficient and gregarious daughter, and lost the baby who really needs her mother in order to enjoy herself.

Lynne, ten years old, is invited to Naomi's birthday party, where they are planning to watch a video her mother, Ms T, thinks unsuitable. Ms T does not want her to see it and wonders whether she should not allow her to go or whether she should speak to Naomi's parents and ask if it would be possible for them to choose a different video. Lynne is desperate that her mother say nothing; it would embarrass her too much. Ms T knows Lynne is terrified of horror films and just does not know what to do. Lynne so much wants to be part of the group that her mother lets her go and says nothing. For weeks after, Lynne wakes up at night with terrible nightmares. Ms T wishes she had stood firm and trusted her instinct about her child.

It is so hard to get it right, we are bound to misread situations and make incorrect judgements. We tend on the whole to be reluctant to say no if this isolates our child, yet at times we need to make the stand. The price paid here is not too high, and Lynne may have learnt that if she wants to do the same as the others, it may not suit her. It is useful to remember that saying no where others say yes gives the child a model that being different is OK. This will help them to resist peer pressure later, on their own. It can assist them in saying no to sex or drugs when it is thought really cool to indulge. You are giving them the idea that they can think, 'Is this really what I want *for me*?' The stakes of not standing up for yourself can be very great.

Our Response to Conflicts

Your child is away from you for at least seven hours a day during the week. When you ask about what happened at school today, you are likely to be faced with a short reply, often just, 'Nothing much.' Even if the child speaks about school a lot, it is not the same as knowing for yourself. In the past you have probably had some contact with his caretaker, whether this be a child-minder, a nanny, an au-pair, a relative or a teacher at his nursery. For the first time you may have to rely totally on his account to get a picture of how he has been. And the question arises of how you deal with difficulties reported back to you – ones you have not witnessed. Your child reports that teacher X hates him and picks on him all the time, or that Sharon always makes fun of him. He may beg you to go in and tell them off or stop it in some way. On the other hand he may plead for you not to mention it at all. Getting as upset as he is or wanting to act out the same impulses may not be the most useful way. Ringing up the school and complaining about Sharon may just inflame the situation. When we respond to what our children tell us, we are giving them a model about how people interact and how to deal with conflict. We can also help them differentiate between feeling and acting.

Sasha, seven years old, came home in tears because she had been told off by the teacher for fighting. She explained to her mother that Rebecca, the little girl who sits next to her, had been teasing her all morning, hiding her pencils and her rubber. Eventually Sasha hit Rebecca with her ruler, which the teacher saw and punished her for. Sasha protested but was told quite firmly that she was not to hit. Her mother acknowledged how unfair it seemed that Sasha had been singled out. She did, however, point out that Sasha had hit and that this was not allowed. They discussed together other ways Sasha might have dealt with Rebecca. Sasha had to think of finding alternative, permissible, means of dealing with her feelings, without acting them out.

As we saw in Chapter I, the issue of translating a difficult or painful experience is also important. If we over-react, we will be giving the message that the outside world is truly dangerous and full of potential enemies. Many families seen in Child Guidance Clinics function on the basis of only being safe within the family and at home. They see all outsiders as unhelpful. This can set up a real problem in taking in good things from others, be they teachers, friends or even books. In an extreme form it can lead to school refusal. Children who are worried about going into school need their parents to believe in the school, to be positive that it has good things to offer and is a safe place to be. If parents are overwhelmed by their child's negative feelings about it, they will not be in a position to give him an alternative view. By becoming as worried as their child, they are in effect agreeing to his version of life at school and not giving him hope that it can be different.

However, not really listening to the child's distress can also lead to problems. It may well be that Sharon, the alleged bully, needs to be tackled, or that teacher X has real difficulty with your son. The idea here is to listen, not just to say yes and accept the report wholesale.

LISTENING

You may find hearing about your child's experience so painful it makes you angry – angry at those who cause him pain, but maybe also at him for exposing you to it. So you may get cross, tell him off for whinging, tell him to pull himself together, tell him he doesn't know what distress is. All these comments may have their place. Our concern here is to know where they are coming from and so to give ourselves time to think and act out of choice, not distress.

In Chapters 1 and 2 we saw how, when somebody is upset, their feelings often get right inside us. We start to feel equally upset. This often revives our own upsets, past and present. If you hear your child's account as if it is happening to you, you will not have the necessary distance to examine the situation carefully. Memories of times you found difficult or painful may flood back, old emotions and battles get revived. Then what could just have been an awkward moment for him gets blown up into a trauma. We have to be watchful that we don't let our children's pains become overwhelming for us. By really listening and taking on board how they feel, but staying detached enough to make sense, we can detoxify a bad experience, give back a different point of view. This can help the child see the event as an obstacle he can overcome. If we cannot manage this, sometimes we add to the child's confusion and worry.

Mr L had been a small, rather puny-looking boy, whom the bigger children picked on and teased. At the time, he had felt victimized and been unable to stand up to the bullies. He had spoken to his parents and teachers about it, but not in a forceful way, afraid of the retribution he had been threatened with. When he heard his son Mark, aged seven, complaining about Roger hitting him during break, Mr L got very angry. He knew Roger's parents and wanted to ring them up, to complain about their son's behaviour. Mark pleaded with him not to and tried to explain that Roger had not meant to hit him. Mr L was being driven to action by the memory of his own childhood fear of 'telling' and assumed Mark was in the same position. He was rather cross with Mark, too, for not letting him deal with the bully. He was reliving his own experience and trying to offer Mark the protection he felt he had not had, the strong protective daddy he had longed for. He also wanted to teach the bullies of the past a lesson by confronting the new one. However, he was persuaded not to intervene. A few days later when the two families were out together, Mr L brought up the subject of Roger hitting Mark. Roger's parents recounted what they had heard, and it seemed the boys had got into a fight, both hitting each other. They had resolved this the next day and were back to being friends.

It would not have been helpful for Mr L to call Roger's parents in anger. He would have got into a fight with his friends, just as Mark had done with Roger. Although he was unaware of this, hearing about his son being hurt triggered the feelings he had when he was Mark's age. At that moment, emotionally, he stopped being a parent and his feelings were those of his childhood. It was hard for him to recognize Mark's experience as different, and respond to what Mark needed. Mark was not asking for help with an external problem, with a bully, but for a place to complain about having fought with a friend, which was what was really troubling him.

The reverse scenario could be as true. A parent who managed, despite adversity, may not be able to hear the real cry for help from a child who is being bullied. The idea that 'I coped whatever the situation' may make him deaf to the fact that his child is not coping.

Other factors may interfere with our capacity to hear the child's experience.

Mr and Mrs S settled in England as refugees. They were both professionals in their homeland but have had to struggle to make ends meet here. The children have adapted well and are successful at school. However, Mary aged nine, reports being badly teased at school. Other children pull her hair, call her names and generally torment her. She is becoming frightened of going to class. The family are very upset by this but feel helpless. They do not intervene and leave Mary to cope on her own.

In this case, Mary's distress reminds her parents of the far more extreme fear they felt before leaving their country, when they never knew if they might be caught and tortured. The memory disturbs them and they quickly wish to get away from it. Added to this, they feel their status in this country is precarious. They have many dealings with the local council and various authority figures. They fear that going to the school to complain would jeopardize their position. They do not want to be seen as trouble-makers. Their dependence on, as well as gratitude to, the system makes it hard for them to criticize the school. They also wish for things just to be normal and are not open to any ripples that might rock the boat. What then happens is that they get irritated with Mary, think she must be provoking the other children and want her to sort it out herself. They find it unbearable to hear what she says and react in a way that inhibits her from telling them anymore. Their feelings are very understandable, though deeply unhelpful to Mary. Sadly, the solutions we find for our difficulties cannot always be the best for everyone in the family.

In all these cases, the important point is to listen to the child as well as to be alert to what is sparked off in us. We have to try to sort out what our own feelings are and what comes from our child in order to know how best to help. It is not easy to see and examine what, emotionally, belongs to us.

STAYING WITH THE PRESENT

In looking at what gets stirred up in us by our children, we are faced with a more complicated issue than is apparent at first. We have seen what a no may mean to your child; it may stand for much more than the simple limit or divergent opinion you mean it to be. Similarly, we frequently react to our child's behaviour in relation to our own point of view. Events have a meaning beyond what is happening in the here and now.

Perhaps a new concern for you is the impact the child has on the group, how he behaves without you, for extended periods of time. A common worry is not just how the child copes, but also how this reflects on you. We have all seen or experienced a meal with a child eating noisily, belching, pulling faces, rocking back and forth on his chair, interrupting conversations, getting up and running around. Rather than dealing with this as it occurs, usually the parent's mind leaps into hyper-drive, imagining all sorts of appalling scenarios. And so the dialogue, or rather harangue, begins: 'I hope you don't behave like this outside. What will people think? How can you possibly learn anything at school if you are so easily distracted? We have always taught you to have good manners, has nothing sunk in? If you keep leaning on that chair you'll fall and hurt yourself and that will teach you!' The child is no longer being reprimanded for poor behaviour; rather it seems as if his whole future depends on this moment. The issue is not what is happening but what will happen: will this behaviour go on for ever, will the child ever manage to be civilized? Alongside this, it is as if one's entire parenting comes into question. Have we failed so miserably?

I am using a trivial example to point out what gets triggered in such conflicts, which are numerous during an average household's day and can take various forms, but the threads are consistent. Our reaction is not helpful, it stops us from seeing what is actually going on, and sends us spinning into the future. We then respond badly to the present moment. How could we deal better with situations like these? One way is to stay in touch with what is happening now with the child, to be aware of what this raises in us, and to try to keep the two separate. This helps us retain a sense of proportion and so search for an appropriate solution.

GUILT

Much of our difficulty in saying no stems from guilt. We overcompensate for what we feel we have deprived our child of.

Mrs K has two children, Adam, aged five and Natasha, three. She gave up work when her children were born but recently her husband needed help with his business and she has taken this on for three months. In the morning she has to rush to school with Adam, then dash to Natasha's nursery and finally commute an hour to work. Mrs K describes poignantly how painful it is leaving Adam all alone in the playground, as he is always the first one there. He has just started at primary school and feels very much a new boy. She is haunted by the image of this small lone child, under the shelter in a huge playground.

She has arranged for the children to be collected at the end of the day by a child-minder who has known them since they were born. She manages to get back at the same time as them. She has planned and arranged everything so that the children suffer the least possible disruption. In fact she is there at either end of the day for them and they are all away at the same time. However, despite all this, she carries a sense of guilt at being so busy. She enjoys her job tremendously once she gets over the hectic start to the morning. But the pressure to get everywhere on time, the lack of flexibility, the fear that the children may be ill, make her feel sick. When talking about it, she uses words like 'guilt', 'fear', 'worry'. When she gets home she overcompensates, trying to make up for not being solely preoccupied with the children. She wants everything between them to be nice and pleasant.

This is many working women's experience. Even those who are not away for longer than the school day feel guilty when they are at home that their mind is on other issues, that they have to do the housework or make phone calls. It is hard not to fall in with the children's expectations that you should be totally available to them when you are together. This often results in the mother believing the child when he tells her she is mean and nasty. She builds a picture of herself as a bad mother and tries to repair the damage she feels she has caused. This often leads to saying yes always and not facing the child's fury and criticisms. This is even more the case for mothers who work full time and need to use extended child care.

Similarly fathers who work long hours and do not see much of their children tend to indulge them. The old-fashioned traditional image of the father as the limit-setter, the disciplinarian, does not hold true today. Few children nowadays hear the famous 'Wait till your father comes home.' It is more often mothers who are responsible for saying no. Whether they are at home all day and worn down by the children's demands or out at work and tired by the various pulls on their time, saying no is not easy. The 'yes' may come out just to hold the peace or to make them feel better about themselves. Unfortunately it is never a long-term answer to the demands of the children or the worry stirred up inside oneself about whether one is good enough.

Guilt interferes in smaller ways, too.

Sophie, aged ten, is doing her homework in front of the television. Her mother asks her to switch it off. 'But it's my favourite programme,' she complains. Mrs M agrees to let her watch but tells her she'll be distracted and that she thinks it is a really bad idea to work like this. Sophie carries on, assuring Mrs M that her work is nearly finished, she's just copying out. When Mrs M checks it, Sophie has made careless mistakes, forgetting words in sentences and writing some twice over. Mrs M is angry with herself for not having stood her ground. She tells Sophie that tomorrow there will be no TV. Sophie cries and tells her she is mean, all the other children watch TV, it's a special episode, a crucial part of the plot, and so on. Mrs M feels cruel and the next day gives in to her daughter's renewed pleading.

These simple conflicts occur in some form in every household. Mrs M did not want an argument with Sophie and left her to her own devices. When this interfered with her homework, Mrs M felt guilty and cross. She reacted by banning TV, a major sanction for her. It is clear that she finds it hard to set firm limits, and Sophie has no problem shifting her mother. In the end there is no limit. Going from no limit to one which seems excessive to the child and yourself is unlikely to work. Mrs M ended up feeling cruel and guilty again. A compromise would have been to let Sophie watch the programme but insist she do her homework at a separate time, with no other distractions. It might have involved an argument and a strict 'no' to her pleas, but it would also have given Mrs M the confidence that Sophie had had one of her requests granted. She would have felt less mean and this would help her be firm.

In cases of divorce or separation, both parents often feel guilty and this can interfere with ordinary limits.

Terry, eight years old, is the only child of Mr and Mrs C, who are separated. Their relationship is amicable and Terry lives with his mother but spends every alternate weekend with his father. He also speaks to him daily on the telephone. They came to see me because Terry had many tantrums, was extremely demanding and would make life a misery if he did not get what he wanted. His father described himself as a softy and could not say no to his son. He also believed that if he were nice to him, he would be nice back. He wanted what was best for Terry and felt very upset that he seemed unhappy. He felt tremendous guilt at not living with him any more, and found it hard not to give in to his every request. Mrs C was also at the end of her tether, feeling all her contact with Terry involved a battle of wills. She was irritated with her ex-husband for indulging Terry, which made it harder for her to be firm. When I saw them together, Terry appeared very pleased with his impact on them. It seemed the main side-effect of his behaviour was that both parents spoke to each other daily about him. He kept their contact constant and highly charged emotionally. It was one way of keeping his parents together, focused on him. His feelings about the loss of his parents as a marital couple were not expressed openly, and maybe not even felt. He insisted everything was fine as far as he was concerned. *He* was not aware of having a problem, *they* were. It was as if one way of trying to resolve his pain was to give them a pain, to fill them with guilt and so stop them from getting on and forming new relationships. He also robbed himself of a proper home with either of them by turning both homes into unhappy places. The parents were very understanding about the impact of their divorce and did what they could to make arrangements for easy access, they lived nearby, and he could telephone either parent at any time. However, their reluctance to be firm allowed him to carry on ruining all their time together.

Sometimes children get so caught up in their own unhappiness or anger they become spoilers, for themselves and others. It is a parental task to stop this before it becomes entrenched. Mr C could not say to his son that he would not take his fifth telephone call in twenty minutes because he got caught up in the fantasy that Terry really needed that many calls. He was not able to stand back and realize that he had just spoken to him. His guilt at hurting Terry, the idea that he had abandoned him in some way, paralysed his thinking. The ordinary limits he would have put on him if he was still at home no longer applied. This lack of limits encouraged Terry to become tyrannical and spoil his time with everyone. In an attempt to avoid themselves or Terry feeling bad, the parents gave in to him. In fact this backfired and they ended up in a constant confrontation with him. Firm limits would have helped Terry realize life had changed and come to terms with the fact that his parents' separation did mean one or other of them would not be as easily available. He would have to mourn the loss. They, too, would have to face his anger about it. The way everybody in the family managed was to try to change as little as possible, to pretend Terry's life could be more or less the same. This is not possible and covers up the truth. Terry's finding fault with everything, demanding constant treats, was his way of filling the gap, avoiding the true impact of not living with both parents. His fear of being rejected was then acted out because in the end each parent was relieved to hand him over to the other.

Guilt also tends to make us feel our children should have many material possessions. Houses now abound with toys, clothes, games. Even the poorest families struggle to give to their children first, as soon as money comes in. A striking feature of families I worked with in a Young Family Centre run by Social Services was how much the mothers spent on their children. As a result children tend to grow up with a sense that goods are disposable, and that a new supply should be constant. They act as if they really 'need' whatever it may be, which taps into our feelings of not giving enough, whether this be material things or time, attention, love. We want to make up for what we feel we have not provided, so we give objects.

Yet by doing this we may deprive the child of a necessary experience. When children want, they feel they need. As adults, we are more able to draw a distinction. It is through our attitude that the child learns to differentiate. This is tremendously useful, or the child will always be at the mercy of desperate wants which can never all be satisfied. Having and discarding easily also robs the child of anything being special. If a toy breaks and is immediately replaced, to spare him pain, he does not get to feel the loss and then recover. This prevents toys gaining emotional meaning and the child gaining depth in his attachments. It also interferes with his sense of reality – the understanding that if you break something, it is damaged and may no longer function.

Another benefit of not always getting what you want, of a parent saying no, is the capacity to bear an empty space, a gap – the issue I return to throughout this book. If all gaps are instantly filled, there is no room for creativity. If a child has a toy for every occasion, he will not use his imagination to cobble something together, to turn one object into another. A box will be a box rather than a potential doll's house, a stick will be confined to the garden rather than becoming a conductor's baton or a gun or whatever captures his fancy. Always having the specific object can keep children very concrete and stunt their capacity to symbolize, to pretend, to invent.

Also, and more importantly, his feeling that a gap is unbearable is reinforced. We are agreeing that not having is terrible, that he is lost without satisfaction. We are in effect saying he is what he has. If a child links his importance to what he possesses, his self-image will always be at risk. In tolerating not having, he is increasing his capacity to believe in himself and the knowledge that it is who he is, his character, that is the most valuable possession of all, and one that he cannot be stripped of. It is this sense of self-worth, of being appreciated for yourself, which helps people survive in times of adversity. In our adult encounters, we may meet many ambitious people who do not have this sense of security.

MIRROR, MIRROR

It is hard to make a stand when you feel you are being unkind. During the primary-school years many parents feel their children really 'get under their skin'. We are of a generation which has been brought up to 'respect' children. We believe this means explaining things to them, but we then expect them to agree! We wish to rule by consensus. This is a form of refusing to acknowledge their difference from us and also their struggle to find their own way of being in the world. Many a seven-year-old will point out his mother's inconsistency in dealing with bad behaviour and make her feel very rapidly that she is unreasonable and definitely at fault. Parents move into a defensive position and thereby lose the argument. It is important for the child's development that he indulge in his rebellion and questioning, that he find his footing and express himself. It is equally important that the adult stick to his own position and not join in the battle of seven-year-olds. In order for that to happen, you have to think before you make a stand – is this something you really believe in? If you are clear, you are much more likely to be firm and not rocked by whatever onslaught may come your way.

I have already described how the child sees himself reflected in his mother's eyes. The corollary is that the mother, too, sees herself reflected in his. At the stage when children are attempting to internalize rules and conventions, they often adopt their parents' mannerisms. They have also learned their parents' view of the world and way of dealing with it. One of the factors which disturbs us, I believe, is seeing aspects of ourselves in our children, particularly the traits we are not so fond of. A child who will argue continuously, will not let a subject drop, or will easily fold in the face of adversity, may well be very like his parent. If this is so, parents are inclined to be harsher on those occasions than they might otherwise be.

We have seen how when we are upset with our children, we frequently regress ourselves, and usually join in at the stage they are at. With babies we panic, with toddlers we have our equivalent of a tantrum, and with the primary-school child we become rather rigid, slightly arrogant and bossy. This does not always happen, but when our children's behaviour really gets to us, it does often take us back to where they are. I was trying to puzzle out what it is about the five to ten age group that seems to arouse our irritation so much, and I think it is to do with their attempts to be like adults, as they see them. This effort remains with us throughout life. Many adults will speak of never having grown up, not properly anyway, as if they have somehow fooled everyone else. It is as if that childhood image of a grown-up stays with us – some kind of super-reasonable, thoughtful, in-control person who is always right. The child who argues, does not listen or comply may threaten our image of ourselves as the adult, and push us into a rigid and stubborn battle for the position. With babies and toddlers, our role is clearly that of parent. With the older child and even more so with teenagers, our position may feel under attack. We get into asking, 'Who is the parent here?' Our confidence is undermined and instead of acting more like parents, that is with our child's and our interests at heart, we become more like pseudo-adults, like the child trying to be the parent.

Sanctions

We have seen that this is an age where logic, justice and fairness play a meaningful part. Children are learning to comply with requirements and need to believe that it is for a reason. It is clear that boundaries and limits are useful, but the question of how to apply them is not so easy. My daughter, Holly, ten years old, is very keen that I should say clearly in my book that it is important for adults to say no kindly, not shout or scold too much, otherwise, she tells me, 'Children just stop listening.' I agree with her. I have talked about limits making a child feel safer, or stretching his emotional muscles. The crucial question is of balance and motivation. If your 'no' is delivered in retaliation or in appeasement, it may not achieve its goal. If it is more to do with your concerns than where your child is at, it may also miss its target. If you believe it is for the good of your child, you will have more conviction and this will be communicated.

Of course, there may be times where you have to enforce your stance, usually with a sanction.

Georgiou, aged six, got very cross when his friend Jim was playing at his house and, in a temper, broke a toy lorry. After his friend had left, Georgiou was very distressed and demanded a new lorry. His mother decided not to replace the toy immediately, so that he realized the impact of his actions. She also told him that he would not be allowed to invite a friend home for one week, as he had behaved so badly.

On the whole if children are asked about sanctions, they tend to expect and to dream up far harsher ones than adults do. This is as it should be. Sanctions are only of true benefit if they foster development. Those which terrorize or force a child into submission do not make for a healthier child. What strengthens a child is a feeling that his parents will put themselves out, spend time thinking about him and be prepared to go through tough times, arguments and battles, for his sake. This means that the sanction that works best is something related to the misdemeanour and in proportion to it, as it was with Georgiou. It is specific and not a global judgement on the whole child. It is strong enough to provoke thought and not so upsetting as to inhibit learning. It is something you believe in and so will stick by, regardless of pressure put on you and whether it makes you unpopular.

It is also important that sanctions should be linked to the child rather than to you. Telling a child that his behaviour makes you ill is an unfair load to place on him, a responsibility which is not his. We can find ourselves saying things like, 'You'll kill me off, if you carry on like that.' This is just not true. He is responsible for his behaviour, you are responsible for how it makes you feel.

We like to think of ourselves as good parents. We expect this to mean that our children think of us as such. When we are in conflict with them, it is hard to keep track of the overall picture. When we say no and they object and accuse us of being horrid, we feel we are the bad parent we see in their eyes or hear in their description of us. This can make it hard for us to say no. We have to be prepared to be unpopular and strong enough in ourselves to hold on to the knowledge that we are not so bad after all.

Summary

During the primary-school years the child's environment shifts from being mainly within the family group to being at school. He has to adapt to being one of a group and fitting in with expectations to a greater extent than previously. His world becomes full of rules and tasks. Saying no is a way of establishing boundaries and separateness, making a space between wishes, thoughts and actions. The whole issue of control, whether it be from within or without, is pertinent to this age group. Primary-school children like rules and structures. They help them to distance themselves from their more infantile feelings, which need to be held in check, so that they are free to concentrate and learn. At this age pleasure is taken in using and exploring language, reason and intellectual skills. The toddler had to learn to walk, run, handle small objects, be dextrous, master language and communication. Between five and ten the child is working on a larger canvas and with finer control. He has to make his own friendships, negotiate conflict, find his place in the social group. For him to be able to do all this, to open himself up to new people, new thoughts, new skills, new learning, he needs to start from a secure base. He has to feel separate, believe in himself and believe the world at large has much to offer. By saying no in the wide and symbolic way I am using the concept, we will have helped him gain a sense of himself, a sense of ourselves, and the capacity to choose how he relates to the world.

4
Adolescents

> I do not consider myself worth counting,
> but sometimes even for me
> heaven and earth are too small.
> KUJO TAKEKO

A Time of Transformation

Historically, adolescence was not recognized by many cultures. As soon as children grew physically mature, girls bore children and boys worked. In our society it isn't clear whether an adolescent is a child, a teenager or a young person. Adolescence also seems to last for longer and longer. In terms of psychotherapy we may even consider someone in their early twenties an adolescent, if they are still dependent on parental figures, such as a student living at home. Of course the world of young adolescents is very different to that of later adolescence, and there are many gender differences too. Psychologists have treated this as a critical period, one where the child returns, after a quieter period of consolidation, to the dramas of early childhood, revisited in the context of a developing sexuality. Adolescence is often spoken of as a hurdle to get over, and caricatured images of teenagers at home always involve trouble. However, it is a period of great developmental opportunity. For parents, too, it can be a very moving and exciting time, to see their young child emerge and grow into an adult.

The issue of saying no gets more complex: should we still set limits and in what manner? It is also a time to impose stricter boundaries on ourselves, in order to promote our children's growth. We have to let go to a greater extent than previously so that they can explore for themselves. The main question in the young person's mind is one of identity: 'Who am I?'

THERE'S A STRANGER IN MY BODY

The early adolescent years, from about twelve to fourteen, are a period of great change. Physical growth is more rapid than at any other time, except in the womb. Sexuality enters the frame. Parents often complain of feeling their child is different, has become a stranger in their home. However, they forget that the child may feel quite a stranger to himself. The body changes dramatically and with this many emotions are stirred up. Hormones are raging, the child may feel tearful, elated, excited; on the whole extremes prevail. There is an enormous preoccupation with the body, its sensations, and with appearances - what he looks like, how he is perceived. Aspects of himself which he has got used to begin to change. Some of these transformations, on which adults invariably comment, are all too apparent. The boy who has just started growing facial hair or the girl who is getting used to having a different shape will frequently be greeted with exclamations of how much they have grown, how they are no longer a child but a young man or woman. An experience which feels very private is visible to all. Girls are confronted not just by physical development, but by the advent of menstruation and the knowledge that they could become mothers. They have to cope with the reality that they are open to adult responsibilities while they are still immature. Boys are less aware of this but carry an anxiety about their potency, often related to conquests or achievement. They also have to deal with the strange experience of their voice breaking, which can really feel like another person suddenly and unpredictably bursting out of them.

Adolescents are often tearful and, if they are close enough to speak to their parents, they will communicate their own bewilderment at the state they are in: 'I'm sad,' 'I feel really lonely,'

'Nobody likes me at school', 'Everybody laughs at me,' followed by, 'But I don't know why.' Even if all the evidence is to the contrary – friends telephoning, the child being admired and his company sought – these feelings of insecurity and isolation are very real.

Another puzzle is how different he can feel from one day to the next: one minute he is on top of the world, another in the depths of despair. His moods and his image of himself sway like branches in the wind. The adolescent feels fragmented, unsure of what to rely on. Will he be happy or totally miserable tomorrow? Similarly the parent doesn't know who to address at any one time. Is Suzy in a confident and cheerful mood so Mum can easily tackle her about the mess in her room, or will it send her into depression? Because the adolescent fluctuates between being independent and quite mature at times and at others being infantile, parents get wrong-footed. If you talk to the young child in him, the adolescent may criticize you for being patronizing, not trusting him. If you treat him as an adult, he may feel pushed and uncared for. This can show itself in the simplest conflicts. For instance, he wants to know what time a film is on and you encourage him to ring the cinema and find out. He accuses you of thinking he is a grown-up, tells you his friends' mothers never ask them to make such calls themselves. Later, when you wish to know who he is going out with and how he will get back, he tells you he is not a baby and not to be so overprotective. At this point it seems as if whichever aspect of him you talk to, you fail to take the other into account. This unpredictability makes everyone in the family feel they are walking on eggshells. The issue of saying no and setting limits becomes delicate and many families feel they are failing at this point.

HOME: A SECURE BASE

During this time of change, of insecurity and movement, our growing child may feel out of control. It is particularly important at this time that we should not be invaded and taken over by the same feelings as him. In Chapter 1 I wrote about a mother's capacity to contain her baby, her ability to transform frightening feelings into more manageable emotions. This function is also crucial with adolescents, but takes a different form. Where the baby was crying in distress, you could just hold and soothe him. The teenager will show his upset quite differently; at times he will be angry, provocative, fearful, sad, confused – a whole gamut of emotions will flood him. At times we will help by talking to him, being there when he is in a crisis. But, more subtly, it is the home setting, the environment we provide for him, that will make him feel held and safe. Our ability to make rules, to stick to them, to have a sense of what is appropriate or not will contribute to how much he feels he can venture forth from a secure base. The key for us is to be strong and flexible.

We may hold our children physically when they feel wobbly, but more than that we hold them in mind. This is one of the ways in which we give them confidence. Parents have to make the move to accommodating new and different aspects of their child. They have to re-adjust their picture of who he is, as he adjusts to changes in himself. We have seen throughout this book that we all develop a sense of who we are partly by looking at our reflection in others' eyes. We know ourselves better by how people react to us, by the emotions evoked in them. If we are different ourselves from day to day, it becomes very hard to get a clear feedback. Parents are faced with the dilemma and onerous task of being open to a new child, as well as being responsible for keeping the child they do know in mind.

It can be a great source of stability for the adolescent to know that his parents feel confident in him, trust that the child they have known up to now is still there and will remain within the new developing person. The important contribution we, as parents, can make is to be welcoming of our children's search for identity, and the many guises they may take till they find what suits them, secure in the knowledge that what is at the core of their personality is good. It is hard to believe this when your teenager seems rebellious, dirty, anti-social or whatever. However, if this positive vision of himself is what he sees reflected in your eyes, it will boost his self-esteem and help him to make wise choices. Of course, I am not talking of blindness to problems and difficulties, nor to a blackmailing stance which states that if you trust him, he cannot let you down. Here I am stressing a basic faith in your child which comes from confidence that you have done your best for him, and the time has come for him to start venturing out on his own.

Reasonable Limits

We have seen in previous chapters how structure, rules and boundaries make children feel safe. During adolescence, rules are often fought against, limits considered frustrating and even at times crippling. Does this mean we should give them up? The adolescent needs to fly, to be the one to break the rules. Again we are faced with a balancing act. The need is two-fold. Firstly, the young adolescent needs parents to struggle against, to have the row with. Just as the baby may need to kick against your hand to get a measure of his force and how far he can stretch, so the adolescent needs a degree of resistance to explore his reach. It is important to allow that and not to try too hard to be the 'good' parent when what he wants is to fight the 'bad' parent he experiences you to be. He may argue with you as a way of finding out what he really thinks; he may reject your point of view in order to look for his own way. Insisting that your children agree with you, or recognize that you are on their side, does not help them venture out into the world. Having a conflict and resolving it will build up their strength.

Secondly, of course, there are times when you need to say no firmly. Sometimes the child really wants you to restrict him; he is frightened or worried about something but does not want to lose face in front of others or be disappointed in his own image of himself as being the adventurous one.

Shabana, twelve, had been asked by her school friends to come shopping with them to a Sunday market in a rather rough part of town. She was excited at the prospect but also nervous. The more she asked her mother if she could go, the clearer it became that her own feelings were very mixed. She wanted to be enthusiastic about the trip but actually she was fearful. Her mother did not give her permission to go. Shabana was upset with her and accused her of being overprotective and unfair. However, on the day of the outing she was very cheerful and warm towards her mother, clearly relieved to have been able to tell her friends she would have loved to go but her mother said no.

In such an ordinary situation the adolescent can maintain the position of being the daring one and you play the role of restricting him, which secretly (occasionally openly) he wants and needs. The same would apply to daily situations where you might encourage your teenager not to stay up too late when you can see he is exhausted, or when your young teenager wants to go out with much older friends and you insist that your normal rules still stand for him, even though they may not for his companions.

The adolescent will speak with passion, will challenge you strongly. It is important for him to hear an echo of his emotions in your voice. If you reply very calmly, or conversely take the pitch much higher, he may feel you have not really heard what he said. He may not know he has reached you. It is important that we get ruffled by their defiance, that we, too, are moved by their emotions. So what matters is how we deal with the questioning of the rules as well as how we voice them.

Our own stand gets wobbly because we are not always sure about what is right for him. It is in these grey areas that we shift our ground. At those times, it will matter to him that you are thinking about what he is saying and that you, too, are struggling with a dilemma. This is helpful to him; it makes him feel you are open and not rigid. What is crucial is to be flexible when we are uncertain, rather than give in because we wish to avoid conflict or cannot bear to be unpopular.

At this age, on the whole, what we can offer mostly is our opinion. Adolescents can disobey and challenge us in ways younger children cannot. They can walk out of the house, they can hurt themselves or you. Rules are mainly enforced because they have been taken in, they exert an internal pressure on him. They are also reinforced by mutual regard. A child who feels his parents care is more likely to respect their decision, and similarly parents who feel their child is open and honest will listen to his point of view. However, there will be times when rules have to be enforced, children not allowed out, privileges withdrawn. As a last resort, help outside the family may have to be called in.

How do adolescents themselves view the issue of setting limits? I spoke with many young teenagers about this book and was surprised how clear they were that they expected and wanted their parents to establish rules. I believe their opinions are quite representative.

- They wanted their parents to stop them smoking and drinking, saying simply, 'Well, they are bad for you, aren't they?' According to this view parents are meant to protect you from what is 'bad'. This did not mean the children would comply with the parents' wishes, but it was the stand they hoped parents would take. They also felt many rules were sensible. Parents who insist on certain information are making a statement about their child. For instance, if you wish to know where he is and when he will be back, you are giving him freedom within a setting. You are saying that boundaries, of time or place, matter. The reason for this is usually to do with ensuring he is safe or can be reached in case of emergency. These concerns give children the feeling they are cared about. You may say no to an outing if these conditions are not met. They may well fight you, accuse you of not trusting them, and so on. You have to judge whether that is true, or whether you are making a stand for their well being as well as your peace of mind. Sticking to your requirement despite their protest and pressure gives them the feeling that you are prepared to put up a fight for their welfare, that you will not let them or others put them at risk. This is a tremendous boost to basic feelings of self-esteem and security.

- They were also very keen that parents should trust them and be open to being convinced if they could present a good argument; that is, they valued flexibility. Here again, availability and listening give the teenager the sense that you are thinking about him, specifically. You are not just applying arbitrary rules. Your 'no' means you are concerned for him. Knowing that you may change your mind means you will be subject to great pressure. Often parents stick to a rigid 'no' so as to avoid a long-drawn-out conflict. However, if you do listen, then your final decision, be it a 'yes' or a 'no', will be related to what has gone on between you. You may end up feeling rotten and your children may think you are awful. In reality, though, you will have given thought to what they said and made up your own mind. This is a good example for them, too: considering all factors, you can take a stand even if this makes you unpopular.
- They believed in establishing the rules early, so that when you get to adolescence, you don't have to struggle with them. Examples included doing homework, helping at home. They warned of the dangers of bribery: 'If you pay someone to do something, they'll want it every time!' They spoke of the struggle to differentiate between when a 'no' is for their benefit and when it is a question of trust. They agreed that at the time a 'no' might often feel like a lack of trust, or provoke a feeling of being unloved, or of unfairness in relation to an older sibling. They were very keen on limits that are the norm and felt parents should check out with each other what is generally allowed. They spoke of how painful it is for those whose parents are overprotective and who miss out on being part of the crowd. They spoke of how hard it was to say 'no' yourself to others, especially those who are particularly popular, those you wish to impress. It seemed easier to say no to strangers than to people you like, or who are still forming an opinion about you.

The adolescents' sensible and thoughtful answers to the question of limits are such a contrast to the experience many parents and children have in the heat of the moment, in the middle of a row. But it served to remind me that adolescence is not just the crisis so often portrayed, a time when all rules and boundaries are attacked. It is also the time when new codes of ethics are formed, when they are working out what sort of adult, and eventually what kind of parent, they want to be, what life they want to live. It is a time of great opportunity.

BEING FIRM

Although rules and boundaries are recognized as important by parents and teenagers, sometimes it is hard to stay firm.

Mr and Mrs P, a middle-class professional couple, have two daughters, aged sixteen and fifteen. Mr P was often away for work abroad and the children's upbringing was mainly left to Mrs P. Mrs P came from a very strict family and grew up quite frightened of her father. When the children were small, she repeated the pattern she was used to, having a very firm regime with them. The children were in bed by 6 p.m., the house was always immaculate, hardly any toys strewn around. Another mother in the children's playgroup commented that 'they make other children look like jungle animals'. Mrs P treated the children much as she was treated, even though she had not enjoyed her childhood. She felt she knew of no other way.

As the children mixed with others, they and Mrs P began to get very different images from their peers. Mrs P started to doubt herself; after all, she hadn't liked her experience. She then attempted to do as she saw others doing, becoming lenient and, in particular, giving up smacking. However, she felt she was left with nothing to replace it with, she had no other means of setting limits. This was manageable while the children were quite small. They were both bright and strong characters, and as they grew older they used all her arguments against her and enormous rows occurred with each one shouting for her point of view. There was little sense of hierarchy or respect; the previous obedience had been based on strength and fear, and now they were all equal and equally distressed.

The situation became quite intolerable when the eldest, Jessica, was sixteen and Mrs P could no longer bear the fights. Jessica achieved good results in her GCSE exams, to everybody's surprise, and insisted on leaving home to take up an apprenticeship. They had many arguments about the pros and cons of such a choice. Both parents agreed this was not a good idea, that she was too young to manage by herself, would lose her friends from school and would be quite isolated among a group likely to be much older than her. They were also anxious about the other risks of being with older people, especially men, rather than the boys she was used to. Jessica argued her case not in terms of what she might gain by going on her own but by criticizing where she was, how she was treated, and complaining how unhappy she was at home.

Although the parents knew this was not the best solution, they allowed her to leave, primarily for peace of mind, for their own sanity as much as anything else. They also told themselves that it was her life and she could do what she liked. Now, as could be predicted, Jessica is very unhappy with her choice and upset that her parents did not stick up for her welfare, that they did not stand their ground and say no to her request and impose on her some more time at home as a child of the family, who was not yet ready for adult responsibilities.

This rather dramatic case illustrates the pitfalls of not having a strong sense of reasonable limits from the beginning. Mrs P had little experience in her childhood of limits being firm but kind. She did not really think of boundaries as helpful so much as rules that have to be obeyed. She therefore had little inner sense of doing something good for her children by being firm. Her own isolation in relation to a busy, absent husband also made her more rigid and afraid of chaos at home. The children in turn felt she feared them and would often accuse their mother of not loving them. However, as is often the case, this feeling would then be transformed from what was genuine upset on their part into a weapon to be used against their mother. When Mrs P felt she could not behave as her parents had toward her, she was left bereft of resources and very much in the position of a child struggling to understand how to manage. There was little scope then for the children's own distress to be contained by an adult. The situation at home became like three children together. At worst this leaves the children feeling parentless. A firm stand with Jessica, motivated by what would be in her best interests, would support her position as a growing child, an adolescent, not yet adult. Unfortunately she is now like a pseudo-adult, prematurely thrown into the world.

In order to set firm, thoughtful limits, we have to have some experience of them as helpful. Our behaviour then comes from an inner confidence which communicates itself to our children.

THE PARENT'S ROLE

If parents do not play the role of parents, it often shows up in problems at school as well as at home.

Hari, twelve, an intelligent boy, had behaviour problems at school and found it hard to manage his anger at home – he would slam doors, shout and disobey. At school he was always in trouble. In sessions he spoke of being bullied at school but had not told his parents, because he felt they would not be able to deal with it. He did not see them as able to act efficiently. He was much brighter than his parents and they often went along with his wishes rather than have a confrontation. When they tried to establish a hierarchy, he would fight and shout them down, so they gave in and felt helpless. He often spoke for the family in meetings and the therapist had to point out to him that she was speaking to his parents. He was not in the position of a child in the home; he knew all the details about the family business, their debts, their regular payments and so on.

Although this may seem like a privileged and powerful position, it was too much for him to handle. Like the little toddler bullies discussed in Chapter 2, it was frightening to be in

Hari's shoes. He felt in charge without the capacity to carry out the task. Therapy tried to redress the balance and re-establish his mother and father in their parental role. Hari had to give up his control but was then able to be a child and develop at a healthier pace.

If teenagers carry too much responsibility, this can get in the way of their growth and can show itself through physical symptoms.

David, fourteen, had 'funny turns', dizziness and terrible headaches. He was a thoughtful, articulate boy who lived in a rough neighbourhood. His mother behaved very much as an adolescent, had three children from different relationships, and moved house regularly without giving much thought to what the changes meant for her children, leaving school and friends behind. There was no regular father figure in their life, although the mother was usually involved with somebody. David took it upon himself to be the little man of the house, was fiercely protective of his siblings, and felt in charge of keeping some regularity in the home. He found it hard to say no to his mother and took on many of her responsibilities. His headaches meant he could get looked after, something he could not ask for emotionally. It was as if being ill was the only way he could have some nurturing. We thought it would be best to talk to his mother about this dynamic and try to shift the load off David. Unfortunately, it was not possible to help the mother in this case, as she refused treatment, feeling fine in herself. She could not see that David's problem had anything to do with feelings or with her. So David was cared for medically when the strain got too much. The symptoms persisted.

There are times when emotions can find no other voice than through the body. Here a parent's infantile needs and her inability to take care of her young family put the burden of adult responsibility on her young son. In this case it is the parent who cannot say no to herself. At other times the child may push to take a parental role, for instance in the absence of a father. In such cases a mother may have to be very firm that he is not to take that on, repeatedly saying no to him playing an inappropriate role. This is not easy and takes courage.

Sometimes we feel so close to our children and their experience that we cannot differentiate between our feelings and theirs.

Neil, eleven years old, was referred to the Child and Adolescent Psychiatry Department because he was finding the transition to secondary school very difficult and had started refusing to go to school. He is the oldest of four children and therefore has not experienced any of his siblings going through the change from small primary group to big senior school happily. Both parents have suffered from depression and are rather mild people who avoid confrontations. The mother is very focused on her job and the father spends a lot of time with the children; he works nights. The father behaves rather like a big kid, playing with his children. They are trying to encourage Neil to grow up, have given him his own keys to the house, but he is not keen on getting to and from school alone. He is very clingy to Dad and is beginning to refuse to go to school. When pushed, he collapses into a heap, has tantrums like a two-year-old. When at school, he spends time in the Special Needs room, even though he does not need extra help. He is showing in every way possible his reluctance to grow up. His parents worry about him a lot and tend to reinforce the idea that the outside world is rather dangerous. They are overprotective of him, treating him as a younger child when he behaves like one.

The therapist who saw them spent time helping the parents to be stronger, firmer. She discussed with them ways of making school and growing up more attractive. She worked on separating out their fears and anxieties from Neil's. His father and mother also spoke to Neil's class teacher and got a picture of a caring and thoughtful school. Here again, we see how school, the new environment, can be presented as appealing or as unsafe, according to the parents' own fantasies or memories. Checking out the reality of the situation for your child, rather than remembering what it was like for you, is crucial. With their confidence increased, they were able to encourage Neil back to school. They got him a bike to ride there and he generally felt more competent and able to cope.

In our wish always to be close to our children, we may rarely take a firm stand, and we may give up our 'no' under pressure. This stems from a fear that having a conflict means a lack of closeness, a breach of your special link. It is as if having the fight, being angry and upset, is too much. But the parent who always says yes then feels resentful. The adolescent, too, feels dissatisfied; he feels it has cost his parent too much. If his parent does not say 'no' or does not stick to it, he ends up feeling like a bully or a nag. Older teenagers, in particular, speak of knowing what their parents think and at their guilt for taking a different stance. They do not wish to upset their parents but may resent feeling guilty. The underlying feelings of exploitation, of manipulation and emotional blackmail are far more undermining than a straightforward row. The avoidance of conflict prevents us from experiencing that disagreements can be resolved. The spectre of an argument looms forever large if it is never had and recovered from.

Parents may be reluctant to say no because they want to see themselves as kind. If the adolescent pushes too much, though, they end up feeling taken advantage of rather than generous. You may also end up with an infantilized child, who thinks he can do everything easily because you've always been there. There was a funny scene in the American TV series *Friends*, when Rachel, a twenty-something, talks of leaving home and managing on her own. She moves in with friends and when she returns laden with shopping they ask how she is paying for it. She waves her credit card triumphantly, until she is forced to recognize that her father will pay the bills.

At present there is a generation of older teens and young people in their early twenties who are still very reliant on their parents. Some of this is due to unemployment and the problems of finding accommodation. But I think it is also linked to parents' inability to say no to the children remaining dependent. By looking after them, they may be depriving them of having to struggle and find ways of coping by themselves.

Nadia, fifteen, is the teenage daughter of Mrs M. Nadia's parents are divorced and she has spent her teenage years travelling across continents from one parent to the other. Both parents are remarried and attempting to create a new family. My colleague and I heard about her while offering marital therapy to her mother and her new husband. They came to us with concerns about a growing distance between them and the difficulties they were having agreeing how to treat this child from a previous marriage. There was much blaming of each other for all the problems with Nadia, particularly in the area of boundaries and rules. Mr M's main complaint was that Mrs M felt so guilty about the break-up of her previous marriage that she indulged Nadia endlessly. Nadia, by her very disturbed and disturbing behaviour, dominated the family. She stole, stayed out late, lied and was generally delinquent. Although we heard numerous accounts of how awful she was, we were struck by how frightening and distressing it must be for her not to feel she really had a place with them.

Many adolescents test out your limits: how much will you put up with before you get rid of them? This is more pronounced in deprived children, children who have been fostered and moved around from home to home, although it is present to some extent in every child. It is a way of asking, 'How much do you really love me?'

Nadia's bravado was very superficial and the upset was not deeply hidden. Once we were able to discuss with the couple their resentment at having to deal with a child from the previous marriage, and their wish at times that they did not have Nadia with them, we were also able to discuss the fact that her behaviour was not just about delinquency. They became aware that there was some truth in Nadia's feeling she had no place with them; they did feel angry at her presence, they did at times want to get rid of her as someone who came in between them. In that way they were able to own some of the feelings themselves and think more carefully about how their new family could integrate the previous one, and what role Mr M would take.

As we disentangled their feelings about starting a new family within an existing one, they were more able to discuss the issue of responsibility for boundaries and rules. Would everything to do with Nadia be left to Mrs M, or would they make a stand that this was the new family with its own rules? In their behaviour and their conversations they became more able to let Nadia know that of course she would always be Mrs M's daughter and have a role in the family, but so would Mr M. As work progressed they insisted on clear communication with Nadia's father regarding her welfare and thought out decisions about how and when she would go to the other parent. The aim was that the visits should become part of an arrangement agreed by all parties, rather than crisis management, as in the past, with the child running away from one to the other, or one or other parent needing a break.

The issue of the adolescent's place in the family is constantly being played out. Is he a lodger in a boarding house? Is he a child to the parents? Is he an older sibling in charge of the younger ones? Is he there to protect or act as partner to a parent? In most families these questions will arise and may often turn into a battleground. Parents have to be parents, the child has to feel secure they are there to stay, to be his parents however he behaves. To be parents, they have to feel in charge of their own home and take responsibility for the kind of family they will have. A child who feels the limits or boundaries are so fragile that he can knock them over will feel insecure and frightened. It encourages his destructiveness and not his creativity. The parents' ability to say no and be firm is a safety net for the scared teenager.

A Separate Identity

Adolescence is a time for trying out an independent identity. The family home is no longer the standard bearer, parents are no longer the people the children wish to emulate. Parents do not hold the central position as they used to do; school and friends replace them as the child's greatest preoccupation. In trying to develop independence, it is important that teenagers move away from their past relationship with their parents. It is very difficult to modify your behaviour and way of thinking with the same people; you need to look elsewhere to practice new skills. Even as adults, we are reluctant to engage in a different type of relationship with our parents. We may behave with them much as we have for most of our life, even if we act quite differently with other people.

So in adolescence there is a search for models outside the family against which teenagers can measure themselves. Ideas and ideologies, religions, systems, fashions and role models are sought. Some of these will be adults – politicians, teachers or authors – some from the world of entertainment, cinema, music, others simply from the children's school and neighbourhood environment. The teenager, for possibly the first time, will be in charge of who he mixes with. He no longer relies on a parent to invite friends over. He will have to negotiate many conflicts by himself; the parent is unlikely to be present as often as before. Adolescents will get together, on public transport, at the cinema, at the local park, out shopping, on a street corner, or whatever. Problems which arise will have to be managed by himself or with his friends. He has to become self-reliant. There is no one there to tell him or remind him of the rules; these have to come from within. By now warnings about dangers or risks or, more benignly, about consideration for others need to have been metabolized by him, to be established. Against this background there may be peer pressure to subscribe to different standards of behaviour.

Whether he chooses to listen to the parental voices from his childhood or not will be up to him at this stage, and we will see that this fluctuates. To help him internalize a sense of boundaries and limits, you have to allow him to choose freely. Parents need to grow and change with their children. They also have to find a new way of being, much as they did when they first had a baby. They now have to let go of their young child and help him to develop into a young adult. Letting go is always hard, throughout life.

Many teenagers need the space to be on their own, to find their place in the peer group by themselves. For some, it is easier to do this by cutting themselves off from the family for a while, because it is too hard in the beginning to be independent while at the same time being close. From the parent's point of view, the child who used to come home from school, want to sit down, be around you, watch television and have a snack and a drink now goes straight to his room and disappears until called. Telephone calls come in constantly or he wants to be out with his friends, and the whole family are aware that he is involved in a social circle which does not include them. Advice is no longer sought about clothes, style, who to see or what to do. Many parents feel terribly excluded.

- Claudia, mother of Julian, twelve and Charlotte, thirteen, found herself at the dinner table between them as they giggled and gossiped about their friends. There were whispers about boyfriends for Charlotte and girls fancying Julian. She could only catch snippets of their conversation and felt a tremendous sense of being left out; there was no space for her in their exciting exchange.
- Paolo, father of fourteen-year-old Tracy, saw her looking distraught and tearful but she would not say what was the matter. She spoke a little to her mother, about an argument she had had with her friends which had upset her, and begged her not to tell Daddy, who might think she was being a baby.
- Richard, fifteen, has always been a quiet child who does not communicate much with his parents. When he was little, he enjoyed just spending time around the family, joining in activities although rarely initiating them. He would also like a cuddle from time to time. As he grew older his retiring nature led him to rather solitary activities such as reading and playing on the computer. Although he seems perfectly content, his parents worry because they feel they have little idea of his emotional life, of who he is growing into. They see other children his age being out and about, possibly in trouble. They worry that Richard is isolated or left out in some way. They wonder whether they should do something.

- Gill, sixteen, spends hours on the phone to her friends yet hardly has anything to say to members of her family. They feel she must have lots on her mind to fill so many conversations, but have no idea what she thinks about or feels.
- Kabir, seventeen, is out every evening and does not tell his parents where he was or who he was with. He always returns when he says he will, but they know nothing of his social life. He is cordial at home and does what is asked of him, helping in the house, taking care of his younger siblings. He is not a problem. Still, his parents feel he is more like a lodger than a son, and worry that they would not know if he was in trouble outside the home. They do not feel he is a part of the family.

These are all everyday examples from ordinary families. Bearing the feeling of being left out can be very hard for parents, but it is crucial to their child's development. The parents have to say no to their wish for closeness and involvement, and may have to settle for just being available if the child wants them and remain sensitive to his feelings and need for privacy.

Although adolescents seem not to value parents' opinion, they are very sensitive to what is said. They may have felt quite secure in their relationship with you, but are wondering if you like the new them, the person they are developing into. For instance, they might think your taste in clothes is really 'sad', but if you criticize a style they think is attractive, they are cut to the quick. My thirteen-year-old daughter and her friends were chatting about buying clothes. They much prefer choosing with each other than with parents. However, one of them was telling me how she had bought a new dress she thought was lovely; when she first wore it, her father looked at her and said, mildly disapproving, 'What's that?' She never wore the dress again.

When adolescents speak with such passion and conviction, we imagine them to be strong and determined. We forget how vulnerable they are at the same time. It is the fluctuating state of their feelings which frequently makes us and them bewildered.

There is, therefore, a constant oscillation from closeness to distance within the family.

Let us try to understand why this distance occurs.

It is very difficult to test out who you are in isolation. On the whole, the young adolescent has a pretty good idea of how he has been perceived at home. He is used to the various one-to-one relationships within the family as well as to belonging to this small nuclear group with quite intense and intimate feelings. At school the situation is different. The primary school crowd he has got used to may well have split up with the move to secondary school. He is faced with a new social circle. He will often form friendships in groups rather than with individuals at this time. The focus shifts from family to social group. Most of his worries and joys are linked to his place within the group. Most of your worries may well be about who he is mixing with. You are no longer the gatekeeper, inviting friends over or visiting his friends' families.

Teenagers' choice of friends is their way of testing the waters. A group may include a quiet child, a noisy one, a funny one, a rebellious one, a sensible one. By being together, they each get to see how people behave towards the others and it is a way of experimenting with different aspects of themselves. They may also do things which you think are very uncharacteristic – steal something from a shop, swear, drink alcohol, smoke, and so on. We have to be prepared for them to taste aspects of life we wish them to avoid. They may need to try out experiences to find out they don't like them, rather than being told, 'Don't do it, it's bad for you.' (I am obviously not talking about a chronic and constant attraction to these activities; I will briefly look at that later.) The choice then carries more conviction and comes from within. A very disciplinarian 'I know better' attitude tends to be counter-productive; it often generates hidden and secretive rebellion, contempt, or a submissiveness which feels like having been bullied. M. Waddell, a child psychotherapist, writes that 'excessively rigid and authoritarian households, who tend to see things in very polarized ways, may inadvertently encourage their children to go to extremes, while believing that what they are doing is setting limits.'

PARENTS ARE ALWAYS WRONG!

Another factor which creates a distance is the child's changing view of his parents. During much of early childhood, children believe their parents are omnipotent, that they know and can do everything. One of our jobs as parents is to disappoint this expectation and give them a more realistic and rounded view. The eminent paediatrician and psychoanalyst, D. W. Winnicott, refers to parents' duty to disillusion their child gradually. In adolescence the child's idealization often turns completely and parents are felt, at times, to be useless. The adolescent feels he and his friends are the only ones to feel passionate and to think strongly about issues. There can be a measure of contempt for the older generation.

One of the normal developmental stages in adolescence is when the teenager thinks that everything the parent (one or both) does is wrong. This is augmented by a tremendous sense of disappointment: how can this same person whom they thought was so fantastic turn out to be such a dope? In order to turn away and separate, it is as if the child needs to feel you are bad. If he continued to think you were wonderful, how would he ever leave you? It is a bit like a weaning again. When feelings are good between two people, there is a wish or dream that somehow this union should go on for ever. But then there would be stasis, no movement, no life.

During adolescence there is a great spurt of growth, things are on the move. So the child turns away, but in order to do so, he may transform you into a bit of a witch or an idiot to make the going away easier. A similar process often occurs with young au-pairs or nannies, prior to departure. Even if up to then you have had a good relationship, you may suddenly find you are having arguments or disagreements about trivial things. These all serve to make leaving less painful. Parents can be quite perplexed when they are accused of being intolerant, small-minded or paranoid by children who just a year ago would look to them for advice. Rows happen, parents and children feel hurt at how they see themselves reflected and thought about. Many parents will engage in long tirades in their own defence and try to convince their offspring of the error of their judgement. This is not very productive; it may stop the teenager making the distance he needs.

If we think of these rows as children's attempts at separation and allow them to express how terrible they feel we are, we are showing that we can tolerate their hatred as well as their love. Being able to have a row in a safe place with someone you know truly loves you can be very reassuring. Winnicott writes of 'the need to defy in a setting in which dependence is met and can be relied upon to be met'.

At such times we need to remember who we really are, and not accept the image presented to us as necessarily accurate. This is akin to saying to ourselves, 'I know he thinks I am nasty, but actually I am not, I am just standing up for what I think is right.' This leads you to carry through your argument, not to give in for fear of upsetting him or of appearing bad in his eyes. We can stick up for our views and enforce our rules. But we cannot compel others to agree that we are right. It is a bit much to ask, as we frequently do, that our teenager not only do what we ask but think it is a good idea too!

LEARNING FROM OUR CHILDREN

It is also true that adults do not have greater knowledge about certain aspects of modern life. Teenagers know more about some new things – developments in information technology, fashion, music, art, slang – than we do. In later adolescence, by thinking for themselves about perennial philosophical issues, they also question values we have grown to take for granted. They push us into re-thinking ethical and political issues. Youth movements have played a great part in major changes, such as the ending of the Vietnam War. Pop musicians, as icons of youth culture, have also played their part in changing attitudes. This was particularly true in the 1960s. For example, the Beatles completely transformed Britain's attitude to Indians. An immigrant people who had been undervalued and thought of as primitive and in need of civilizing was suddenly perceived as holding ancient wisdom coveted by the West. Clothing which had been ridiculed was imitated and integrated. As an Indian myself, this period was extraordinary to behold. A grey country became multi-coloured. The mixing of race and culture often comes from youth's openness to new ways.

Today's young certainly remind us to consider the impact we have on our environment, on how we behave towards animals, how we treat our own bodies. We must not underestimate how much adults can learn from their children; in many ways we look to them to express idealism and hope to counter some of our cynicism.

Sometimes children are also better at academic work or at certain skills or sports than their parents. Here again, recognizing the child's talent, without feeling that it diminishes one's own authority or self-respect, is crucial. Acting as if we are all-knowing about everything is likely to alienate our childen. If we speak knowledgeably about what we know and listen to what they know, then we can have a conversation, a reciprocity, otherwise we slip into a sermon. A great source of irritation for teenagers is to receive a lecture from a parent about something they actually know more about.

Showing that we can learn from them fulfils three functions (at least!). Firstly, it shows them that they have something worthwhile to contribute. This boosts their self-esteem as well as satisfying their wish to give something in return for all they have received. Second, it sets the model that learning is never over, it is always open to renewal and extension. This should give them an open and searching attitude to things that come their way. Third, at a time when they are surging forwards, it reassures them that their parents are not static, not stuck in one place for ever. Some adolescents feel guilty for blossoming and taking charge of themselves. They may worry that their parents are jealous or that somehow there isn't enough liveliness to go round, as if their parents are suddenly very old and will be left behind. The parents' ability to grow and change too gives them a licence to pursue their own development freely.

As adolescents reject their parents or change their perception of their qualities, they suffer a feeling of loss, a sadness at no longer having parents to look up to. They can feel quite bereft and empty for a time. Their own self-esteem suffers, too. If what you have tried to emulate no longer seems so good, you feel yourself diminished. Most children recover from this position and grow to see their parents in the round again, good bits, bad bits and all. It is the struggle with ambivalence that carries on throughout our lives.

TRUE LOVE

Having made a distance between themselves and their parents at a time when emotions are very intense and passionate, in later adolescence teenagers may try to find another relationship in which they feel intimate and close. It is often experienced as a union, a symbiosis, a search for their other half. We have looked at the adolescent's sense of being incomplete, and sometimes finding another who fits you well can make you feel whole. This has elements of the idealized relationship between mother and baby, the 'perfect fit', being a part of each other, which we discussed in chapter I. Classical literature and popular culture are full of examples of the wrench of separation and the yearning to be together always.

Parents may worry that their son or daughter has become deeply involved with somebody and is missing out on group activities or on being part of a wider circle. Although the idealized union and the wish to make the feeling of togetherness last for ever is a necessary developmental stage, we have also seen that it keeps things static, does not allow for movement or growth. If we understand that this kind of relationship is an attempt to recapture a lost dream, we may react to it with less worry and trust that, just as our children grew from their first closeness with us out into the wider world, they may do so again. This will stop us from interfering and trying to say no to their attachment before it has fulfilled its function. We will have to put a limit on ourselves, say no to our wish to spare the child the pain of the broken heart we anticipate. If we do say no to them, we are seen as spoilers – of relationships, of dreams and ideals. Many families have broken links over choices which are considered unsuitable. By setting a limit too soon, you may push the child further away from you, sometimes so far that he does not feel he can return. Here I am not talking about someone whom you think is a danger to your child, just somebody you would not have chosen for him.

Sometimes our objection is linked to our idealization of our child, the view that no one will ever be good enough for him. You may also identify yourself with him, seeing him repeat patterns you fell into at his age. You feel an urge to spare him the pains of your past, and to avoid revisiting them yourself, as you see him suffer. Here again, it is important to make a space, to remember you are different people. Your memory may help you to understand him, but it may also cloud your vision of his particular experience.

As we struggle with the fact that he is different from us, we are sometimes quite unable to see where the attraction lies. How could they choose this person?

Mr and Mrs Z have three children. They are a close family and go on many outings together. The children all go to synagogue and belong to a Jewish Youth Club. Mandy has now fallen in love with a boy she met at a party. He is not Jewish and her parents find this very difficult.

The parents actually like the boy, they find him polite and charming. He is kind to their daughter and respectful to them. Yet they cannot accept Mandy's choice. They keep wondering why she could not find a suitable boy within the community. They wish he could be part of something they know and understand. They would like to be able to place him.

This situation applies to many parents. The community may be religious or geographic, cultural, financial or racial. What is difficult is that the child chooses out of the known, so we have no markers by which to judge. Occasionally it may be just, 'How can he find her attractive?' which translates as, 'How can his taste be so different from mine?' The difference accentuates the fact that you are two separate beings.

Teenagers themselves are aware that when they are in love they behave in ways which feel out of control. They may be teased or tease their friends. Shakespeare wrote: 'I do much wonder that one man, seeing how much another man is a fool when he dedicates his behaviours to love, will, after he hath laughed at such shallow follies in others, become the argument of his own scorn by falling in love.'

Many adults raise their eyebrows in despair or bemusement at their children's antics when in love. It may seem amusing and touching to us, but it is desperately serious to them. We need to remember and accept their feelings without getting swept away by them. A teenager who is distraught about his love life may look to you for help. If we cannot bear to see our child hurt, we may turn against those we feel are causing him pain. We may hate the girlfriend who is behind it all. We may recommend he forget all about her. We speak of other fish in the sea and so on. However, we then act as if the hurt has happened to us, rather than trying to help him think about how he feels. When his pain turns into ours or into our anger, it is hard for him to live through it. The best we can do may be just to stay with him and help him survive it.

SEXUALITY

Adolescents are discovering many new things, especially sexuality. They usually turn away from the thought that parents have a sexual life. This area of the adolescent's life is not for sharing with parents. Worries and concerns may be expressed, but not very fully, because it would make the relationship too intimate and there is a strong taboo not just against incest acted out, but also against incestuous feelings and thoughts. I have often referred to the parents' eyes as mirrors in which the child sees himself reflected. A young person cannot explore his identity as a sexual being with his parents; he has to look elsewhere for this, and may feel very frightened and exposed at having to search for such a crucial part of himself outside the home. This does not mean that the teenager will no longer feel physically close to his parents, and may well still cuddle them. For some, however, the boundary between the physical and the sexual may be too blurred and they may create a distance between themselves and their parents. This is particularly true across the genders, for fathers and daughters, mothers and sons. Here again, it is important for parents to say no to themselves, to tolerate their wish for closeness and so allow the child to find the distance which is comfortable for him.

Physical changes exacerbate the feeling that the teenager does not know who he is or what he will look like eventually; there is a vulnerability born out of not being complete. Boys may not know what to do with facial hair that is not quite a beard. Girls may find a spot on their face which seems an enormous crater to them, which they assume everybody is looking at and which will never disappear; parts of their body grow and they feel they look fat. The French psychoanalyst Françoise Dolto refers to the pains of adolescence as the 'lobster complex', alluding to the time when the lobster sheds his shell and is thoroughly exposed and vulnerable for the time it takes him to grow a new one. They are turning into young adults with the world of sexual encounters opening up to them. One of the ways they get a picture of who they are is linked to how attractive to others they feel. In searching for an image of who they are and who they might be attracted to, they will experiment. Some may turn to homosexual encounters, some may seek many partners. Those things can alarm parents. It is important to notice whether these are transient phases. We must not assume that at a period of such change, anything is for ever. We will have to make an effort to stay in the here and now, and not let our minds leap helter-skelter to the future.

An adolescent who wishes for the security of feeling loved may look to sexual encounters to provide the warmth and affection he seeks. The more infantile need for comfort mingles with the physical ardour of sexuality. Many are vulnerable to exploitation at this time. In the analogy of the lobster, they are frequently tailed by vicious conger eels, waiting to pounce and devour them.

Dee, sixteen years old, was a very delicate and attractive girl. Her parents were separated. She was very close to her mother, who died suddenly. Dee then went to live with the father she hardly knew. The arrangement fell apart and Dee was eventually fostered. She went from being rather overprotected and almost inseparable from her mother to being very much alone and bereft. Her need for a mother figure was palpable in the room during her sessions, and it was very hard for me not to feel drawn into wanting to take her home. She had never even kissed a boy and spoke in a shy and embarrassed way about her curiosity about all things sexual. However, very quickly following her foster placement, she began to get involved sexually with older men, who were clearly taking advantage of her. She spoke of needing to feel somebody loved her, and for the short while they touched her she felt warm and worthwhile.

Many young people look for comfort and an unquestioning love, like the love given by parents, in their sexual relationships. When there is a risk of exploitation, their own capacity to say no can come to the rescue. Teenagers who have a strong sense of being worthwhile and valued have more strength to turn down an encounter which is abusive. Having had the experience of parents who have stuck firm to their boundaries, said no to them in spite of protest and argument, also gives them a model that we can say no to someone and still love them, and likewise we can be told 'no' and still have a close relationship. They realize that a 'no' is a healthy part of a lively relationship rather than its death blow.

The older teenager is in a paradoxical situation when it comes to sexuality. He is legally above the age of consent and yet may not be considered so at home. Attitudes to sexuality have changed very rapidly in the last fifty years and today's sixteen-year-olds bear little resemblance to us at the same age. They may not be emotionally very mature, but their exposure to sexuality is bound to be greater than ours was, and in some cases they may consider their parents naive. One seventeen-year-old pointed out to me the absurdity from her point of view of parents banning their teenager from sleeping over at a friend's because of the worry that it meant she would be having sex there. She said, bemused, 'It is as if they think it has to be later than ten o'clock or in a bedroom!' Older teenagers, particularly girls, tend to appreciate knowing that they can go to parents for advice on sexual relationships. They will also value a parent's concern which shows itself by thinking about problems with them, rather than telling them what to do.

If we want our children to make a stand sexually – to be involved when they want to be, to say no confidently when they want to – in other words to be in charge of their own body, we too have to respect this. There is no point in them doing something or not doing it because their parents want them to. If the choice is in others' hands, anybody can pressurize them. Being secure in your own choice helps withstand pressure not just from individuals but also from the group. When all his friends are talking about their exploits or conquests, a teenager can hold on to his decision and sometimes even say he did not do X or Y because he did not want to.

Some parents are very threatened by their children's emerging sexuality. They may feel old, unattractive, past it. Much of adolescent preening and strutting reinforces this view of parents as no longer sexual. Occasionally, the parent cannot bear to feel left behind and may join in with adolescent crowds or fashions. We see mothers and daughters or fathers and sons out clubbing like siblings. This may be fun from time to time, but is not particularly helpful generally. It does not allow the child to move away nor the parent to let go. It is a pretence that they are the same when they are not. These arrangements are often arrived at in order to appease each other, from a fear of conflict. Both fear the parent's envy. For the budding adult to blossom, he needs to feel he has his parents' blessing, that it gives them joy to see him thrive and grow beautiful. The motivation behind the limits we set will affect how they are received. If we say no out of a wish to spoil, out of jealousy, we may get secrecy and lies; if it comes from a wish to protect, it offers support.

Accepting Difference

Having an adolescent in the home may transform everybody's life. If we welcome him in his new role, we may be invaded by hordes of youngsters, a lot of noise and be expected to feed many hungry mouths. Equally, he may be out for great periods of time and give you little feedback about what he is doing. The question of limits will no doubt cause friction. Whose needs take precedence? If we agree that he needs to be in a group, and that we wish him to be free to do so near us, we have to put up with some inconvenience. What about others in the family? Does the younger child feel terribly isolated seeing his sibling surrounded? Does the noise and mess interfere with the parents' work or leisure time? Who says 'no' to whom? If he is out a lot, how do you bear feeling excluded?

We are back to the balancing act. Being prepared to accommodate the changes in his social life encourages the adolescent to remain close without him feeling closed in or chased out. We also have to be prepared for him to be very different from what he was and different from us. It is part of him finding out about who he wants to be.

Ricky, fifteen, was referred to a child psychotherapist because his parents thought he was 'mad'. They described him as wearing crazy clothes, being secretive, not keeping to house rules, staying out late, not observing school deadlines, being messy . . . When the therapist met him, she expected to see a bit of a monster. It turned out that Ricky looked typically teenage, with bright, slightly eccentric clothes. In speaking to her, he was open and was clearly struggling with his parents' view of him. He felt he was like most of his friends. Meeting the family together, a clearer picture emerged. Ricky was the youngest of three, with a ten-year gap between him and his next sibling. The parents were rather elderly and were both only children. They had been more involved in the social life of their two older children, saw what their friends looked like, how they behaved, and had also chatted with other parents. In talking to them, the therapist was struck by how distant they now felt from their own teenage life and that of their other children. They were no longer in touch with it. They themselves had both been rather compliant and not rebellious in adolescence and just could not understand Ricky. They felt hurt and unappreciated by him. He was also the baby of the family, so to speak, and it was hard for everyone to let him grow up.

With help, Ricky's parents came to see him as wanting privacy, rather than being secretive, and adventurous rather than crazy. They realized they felt threatened by what they experienced as his wildness, more so than if they had been younger or closer to other parents of similar aged children. Although reassured about Ricky, they now had to face the task of adapting to a new age, of taking on their own difficulties rather than explaining their problems in terms of Ricky's madness. Having already got two grown-up children, it was hard to start again. When Ricky was smaller, the other two had also helped parent him; now his parents felt quite alone, worried and resentful. It would have been more comfortable for them to think of it all as Ricky's problem.

Adolescence can stir up many anxieties in adults, fear of wildness, of lack of control. For parents for whom these issues are loaded, an adolescent can be like a cat among pigeons. All hell is let loose. They feel filled with alarm and rather than look inside themselves at what they fear, they see their child as the source of all their worries, as Ricky's parents did. He was made to feel pathological, crazy, for being just an ordinary adolescent. Without the help of his therapist, Ricky might have grown to accept the picture which was evolving of him, and might have ended up actually acting quite mad.

If the child is not allowed to go through his paces, to test out the boundaries at home, he may do this elsewhere.

Jason, fourteen, is his mother's only child. She had him when she was sixteen. He is very aggressive at school, continually suspended for rudeness and generally getting into trouble. At home, Ms A is very strict, saying no all the time. During the week, he is not allowed to watch TV, to go out, has to go to bed early, and so on. Jason is obedient at home, and rebellious at school. In the meetings with her therapist, Ms A spoke of her regret at having Jason so early and of how she had missed out on her adolescence. Now that Jason was old enough, she was going out a lot on her own and enjoying herself. Jason would obey her rules, even when she was out. It seemed he could only be delinquent at school. In joint meetings, Ms A was very surprised to hear how concerned Jason was for her welfare. He felt responsible in many ways for keeping her happy. He could not disobey her or complain because he felt she would not manage. He had accurately perceived that her need for rigid rules was a way of protecting herself against chaos, so he complied. Talking to each other helped and some support was found for Ms A, so as to relieve Jason from part of the burden.

This is quite a common scenario, where the children take on a parental role in the home but then have nowhere suitable to deal with their rebelliousness, anxieties and difficulties. The parents' wish to maintain the status quo, not to change their life too much, hinders the teenager's development. The parents have to say no to their wish to keep things the same and be open to change and dissonance.

The D family came to see me and a colleague because they were concerned that their fifteen-year-old daughter, Sharon, was stealing. They were not wealthy and lived in a very deprived inner-city area. Although the whole family was invited (mother, father, their daughters Tracy, eighteen, and Sharon, fifteen), Mr D only came to see us twice. He was clearly marginalized in the family, which was dominated by Mrs D. Mrs D and Tracy came in looking strikingly similar, with an enormous amount of gold jewellery on their ears and fingers. Their hair was immaculately styled and they wore tight leggings and leather jackets. They were both very talkative and lively and quite overpowering. It was obvious that they spent a lot of time and money on their appearance and clearly enjoyed this together. In contrast Sharon was overweight, scruffy and sullen.

Mrs D and Tracy proceeded to list all Sharon's shortcomings, and session after session was taken up with a litany of her wrongdoings and their exasperation with her. They spoke as one, adding examples to each other's complaints and furnishing details to what sounded like a single point of view. Once we started to probe, however, it became clear that it was very difficult in this household to hold a different standpoint to Mrs D's. Mr D managed by absenting himself, if not physically then emotionally. Tracy, it emerged, spent all her free time with her mother. She had a job, which started very early in the morning, to which her mum drove her, and she was then free from mid-morning to spend the rest of the day at home. She had no interest in friends her own age and preferred to go out window shopping or to the gym with her mum. The accepted family culture was that Mum's way was best.

Sharon rebelled in every way she could, stealing from Mrs D, giving away her sister's clothes, messing up her room, leaving dirty clothes and food under her bed . . . It was as if she could not find another way of separating. She had to be the same or totally opposite. Although at first she seemed the more worrying of the two daughters, we eventually became far more concerned about Tracy. Sharon was behaving in a delinquent manner but was attempting to establish some sort of independence. Her symptoms were directly linked to her relationships within the family. If we could help her stand her ground and find a more acceptable way of being different, we felt she would be OK.

Tracy on the other hand had become entrenched in a whole way of being, more like a twin or a clone of her mother, and showed no wish to change. There was little differentiation between them, as if they were both teenagers or both mothers (to naughty Sharon), and we worried for Tracy's future relationships.

We know that we can also hate those we love, but it may be very difficult to admit this to ourselves. One way of managing these feelings is to split them, so that X can be thought of as wonderful, whereas Y is terrible. This is a common way of trying to keep those we love 'good' and our relationship with them smooth. So with Tracy, being just like Mrs D and idealizing her while rejecting her sister who was so different was a way of keeping close to her mother. In fact as we started to focus gently on Tracy and not join in the haranguing or putting down of Sharon, the family withdrew – Tracy and her mother had too much to lose if they were addressed separately. However, during the short period of treatment we had seen Sharon gain a little confidence, enrol for some classes and get a holiday job.

This case illustrates how difficult it can be to function differently from one parent. The push to conform, to be the same, can be reinforced by cultural and religious norms, too. Not accepting different ways of being is always costly, both to the individual and to the family. A price has to be paid.

Just as the young toddler keeps getting up and falling over again, or the nine-year-old insists on doing his homework with no help at all, even though he is struggling and upset, the teenager needs to stretch himself, to test things out and see what he likes or dislikes, what he wants for himself. He may need to fall and pick himself up, as he did when he was younger. As parents, we may fear his hurt or how he may hurt others. We can offer help as we did with our little ones, but now we cannot restrain him much. Our input is less concrete and physical and more in the nature of offering thoughtfulness.

In the examples of Tracy and of Ricky we see that the parents' view could not encompass the child's variation from them. Such intense resistance to insight implies that elements of the parents' personality are very fragile. Most of us will react in a less extreme fashion, but we are not immune to such reactions. If a father always wears a suit to work, he will not understand why his son goes to an interview without a tie. He may feel irritated that he has had to comply with certain rules all his life, so why should his son get away with less application? Does this mean that his own efforts were for nothing? Or, worse yet, does it mean that what was valid for him is no longer so? Has the value of his experience become redundant? What then transpires between father and son is less to do with whether or not the boy should wear a tie and more to do with self-esteem and feelings of worth.

A child who wants to look and act different can be seen as rejecting and critical. Some may find him provocative or contemptuous. For people who believe that there is little variation in what is acceptable, any divergence is experienced as a threat. Extremes of this stance can be seen in xenophobia, racism, and religious bigotry. If you believe there is only one truth, another point of view is a challenge rather than a contribution. It all depends on where you start from. I remember vividly the difference between my mother's and my grandmother's reaction to young men having long hair in the sixties. My grandmother was alarmed, thinking they looked dirty and that you could not tell boys from girls. My mother thought they looked so much softer and nobler than those with harsh army-style crew cuts; they reminded her of musketeers!

How new fashions, new ways of behaving are received will have an impact on how the adolescent develops. When he adopts a particular new look, language or stance, we are not really meant to like it. It is not tailored to please us, but his peers. If his attempts to hold an individual stance is met by little or no reaction, if we accept it straightaway, he may feel unseen or unheard. It is as if his effort to be outrageous is not noticed. The delicate balance we need to achieve is to act negatively when the aim is to shock without making him feel rejected. He may want us to dislike the style, but it is crucial to like the child. If we rebuff our teenager when he tries out ways of being we would never dream of for ourselves, if we say no to the identities he tries out for himself, we will be doing him a great disservice. A teenager who is consistently rejected for being different will crush his development in order to be accepted, rebel in an extreme way, or look elsewhere for acceptance.

ONE OF THE GANG

A worrying feature of adolescence is many teenagers' pull towards a gang mentality. This is usually born out of a need to belong, to feel safe. The lyrics from *West Side Story* come to mind, describing the gang called the Jets:

When you're a Jet,
You're a Jet all the way,
From your first cigarette,
To your last dyin' day.
When you're a Jet,
If the spit hits the fan,
You got brothers around,
You're a family man.
You're never alone,
You're never disconnected!
You're home with your own:
When company's expected,
You're well protected!

EXCERPT OF JET SONG
(Bernstein/Sondheim)

We can see that the gang is there as a defence, a group to make you feel you have a place, you are always welcome. Gangs often work by excluding others or being against others. The gang member's identity is tied up with making a stand for his companions. This makes each member feel important.

When you're a Jet,
Little boy, you're a man,
Little man, you're a King.

The gang is often a refuge for children who feel they do not have a rightful place or what they need at home. They look for it elsewhere. Unfortunately the gang is not a flexible unit with developmental interests at heart, as a family should be. It is a defensive unit, aiming to protect its members against pain and insecurity. Unlike the social group which I described earlier, where being together is a way of exploring different aspects of oneself, the gang is a rigid structure. Differences are not windows into people's lives which might help you look at and taste other ways of being. They are frightening and strongly resisted. A difference is seen as a challenge and a threat. So you have to be 'in' or you are the enemy.

The gang functions by trying to get rid of distress on to other people, so children who feel they do not have a secure base in their family may deal with this by excluding others, making them feel the pain of insecurity. Children whose experience with their parents is painful, even if the parents do not mean it to be so, protect themselves by slipping into the role of the aggressor. Rather than feel the pain, they inflict it. They are adamant they will not be hurt or feel vulnerable. The gang serves to foster a pseudo-independence, a sense that they are fine and need nobody else, and that all other ways of being are stupid. With this attitude, though, they restrict their world, and are inhibited from learning from benign adults as well as from their peers.

As parents, if we see our children drifting into a gang and it becomes permanent rather than a mild flirtation, we may need to fight for them to return home. This may mean questioning ourselves about what it is at home which is sending them away. Although it is important to encourage exploration, to give children a secure base from which to venture out, adolescence is also a vulnerable time. All major psychiatric illnesses settle during adolescence. It is a time when eating disorders, depressive illnesses, drug and alcohol addiction, and delinquency can become entrenched.

LOSING THEIR WAY

We cannot afford to turn a blind eye to our children being involved in nastiness. As with younger children becoming little tyrants and bullies, it is crucial that parents make a stand, do their best not to let a child behave in ways that are destructive to himself or others. There may be scenes but on the whole children are grateful to parents who support their positive qualities and discourage them from sabotaging their life. This applies to all self-destructive behaviour, such as taking drugs or eating disorders. Parents need to reaffirm their role of being in charge of the family, of setting limits, and, if need be, getting outside help to untangle problems which seem intractable.

It may be difficult to decide what is ordinary rebellion and what is pathology. The teenager may feel out of control for ordinary reasons, as we saw earlier. Much of his behaviour will be aimed at forcing your hand, making you become worried, so that you are firmer. It is not possible within the confines of this book to look very closely at pathology. Generally, however, the adolescent who needs extra help will alarm you deeply. The daughter who comes home with blue hair and a pierced nose but who is her usual cheery (or grumpy!) self will be less worrying than the one who at first sight looks the same but is consistently unhappy, has no sense of humour, no *joie de vivre*.

Most teenagers indulge in a certain amount of delinquency at some time; it could be smoking marijuana, being slightly promiscuous (by this I mean in terms of numbers rather than full sexual encounters), lying to their parents, defying the rules, and so on. These are just the ordinary provocations of adolescence. You need to worry when activities pursued for a brief thrill, or for a sense of fun, take on a grim face, when the teenager seems to want to obliterate feeling or need. Then the pursuit of delinquency has an addictive quality, it appears compulsive and he never seems really to feel better.

If you sense that the child faces catastrophic anxieties, or you experience them, worrying for his health and even his life, as in the case of drug addiction or anorexia, it is essential to seek advice. If you are offering a secure place in your home for your child, if you are clear that you are the parent and he is the child, come what may, yet the worries are unbearable, there is no shame – indeed it should be a mark of courage – in seeking help. You will have to be prepared to think of your role in the constellation too.

FULFILLING OUR DREAMS

The adolescent who is emerging into adulthood often carries his parents' dreams. They may wish to give him what they never had, but also expect him to become what they wanted to be. This is particularly prevalent in the later adolescent years. It becomes most apparent in relation to academic work, choice of subjects, university or career. The young person, if he is strong enough, will remind the parent that it is he who has to do the work not them, and that he will have to live with his choice. Here the child will say no to the parent. This can be hard, and sometimes even as adults we feel the burden of our parents' expectations pushing us on, as we try to give them what they hoped we would provide.

Arish, fifteen, comes from a middle-class family. He is a very bright boy and is expected to achieve high grades at GCSE. He got As in all his mocks. His parents are very ambitious for their children, especially Arish, as he does so well. Arish was referred because he became very depressed, expressed suicidal thoughts, had stopped eating, could not concentrate and was sleeping badly. The family were close and the parents very concerned. He had always been a very 'good' boy, who had done his best and responded to parental demands. He had never been naughty or rebellious. In talking with the family, it emerged that Arish felt under enormous pressure; he thought he was never good enough; whatever he achieved, he felt he was expected to have done better. He spoke of his depression following his grades at mock GCSE when his father asked why he had not got an A star.

His parents were very shocked to hear how much strain he experienced. They had not been aware of it, and from then on they worked enormously hard to control their contribution to this. Arish, too, was encouraged to see his part in the problem, the fact that he had not let them know how he felt. The therapist helped them think of practical ways of going out and having a nice time together, of taking the focus off work for a bit, in order to start their dialogue. In the meetings Arish was able to voice for the first time his anger at the position he felt he held, of being responsible for fulfilling his father's dreams. The parents took all of this on board and looked at the situation differently. Arish also realized that he could speak of his irritation, upset and anger without overwhelming his parents. Soon Arish's depression and suicidal feelings lifted.

We have seen another form of pressure when parents try to select their children's friends and loves. We can get too involved, and need to say no to our wish to determine who matters. If we do not, we can rob our child of finding out for himself if this is a person he really wants to be with. Many parents can get attached to a boyfriend or girlfriend they think eminently suitable, they reinforce the teenager's choice and then become upset when they split up. They may keep in touch with the separate partner and make it difficult for their child to move on and make other choices. Once again, the important question is, whose life is it? We may be unaware of living through our children, we may think we just want what is best for them. There has to be a distance between us, a space to gain some perspective. We should perhaps wait a bit before we offer suggestions, leave a gap in which they can express their wishes.

A Second Chance

As emotions run so high during adolescence, it can be an opportunity to deal with resentments or hurts that have been buried. During the middle years when children want stability and avoid rocking the boat, certain family dynamics get entrenched. With the changes of adolescence, things which seemed blocked can move again.

Maria, aged seventeen, came to see me because she was very depressed. She was at college and finding it hard to study. She had been born with a severe malformation, had had many operations since birth, and was still undergoing occasional surgery. She seemed accustomed to her disability and felt it often embarrassed others more than herself. In our sessions much time was taken up in talking about her relationship with her mother. Maria believed her mother hated her. She had hit her quite severely as a child and Maria had moved in and out of her home, living for some time with her father and for some time with a friend of her mother's. They were both together again and problems were recurring.

Having heard of the early traumas, when Maria spent so much time in hospital, I wondered how this had damaged her relationship with her mother. Generally adolescents are seen on their own in therapy, but I took the unusual step of asking to see the mother and daughter together, as I might with a younger child and parent. As I had suspected, the early years were horrendous for both of them. They had never really spoken of this and both held myths in their minds about how it had felt for the other. They had woven their own story about what the experience meant. It was time to check out reality.

Maria was shocked to hear that her mother had wept at the separation. At the time mothers were only allowed to visit that hospital very briefly. Ms N told us of how she bribed the cleaners to let her hide in the toilet so that she could sneak in and spend extra time with Maria. She spoke of how fearful she had been for Maria's future ability to walk, how she had had to push her in a pushchair for years. The image Maria held of a detached, hateful mother who abandoned her in hospital took a jolt. Similarly Maria, who to her mother had always seemed remote, who did not want to be cuddled or fussed over, spoke of her wish for closeness but her fear of rejection. It was as if for all these years they'd lived with these monstrous images of each other and had never been able to voice them. As they heard each other's experience they started on the road to repair.

It would take a long time and certainly not be easy – too much had happened. But a fundamental shift in their view of each other occurred. There was no doubt that Maria had felt hurt and abandoned, but she saw for the first time that her mother had not wanted to hurt her, and had suffered terribly too.

We see here the possibility of breaking patterns that have been established for years, if the time is right and people are open to change. I cannot help wondering how much difference open hospital visiting would have made to Maria and her mother. The reason why parents were not allowed to visit freely in the past was that they were felt to upset the children, who became more demanding in their presence and protested on their departure. James and Joyce Robertson, who pioneered research into the effects of separation on young children, had a dramatic impact on hospital care for children by showing the detrimental effect of not allowing parents to visit. Their very powerful films made in the 1950s showed that it was harder for staff to deal with the distress when parents left and for the parents themselves to leave a crying child, but for the child, the experience was more bearable if parents could be present. Not given this opportunity, the child may keep his feelings to himself, withdraw and become depressed or act in ways which get rid of the feeling for him, such as hitting another child or damaging his own toys. Dealing with the child's distress is an important part of getting used to separation and preparing the ground for future ones. It shows that difficult experiences can be borne and survived and that you can devise strategies for dealing with them.

In adolescence, too, there are going to be painful aspects to separation and individuation. If you avoid them, cover them up or repress them, you are storing up trouble for later. For instance, a person who never experimented in adolescence may become very envious of their own teenager or identify with them and long to have a wild time. This could lead to having affairs. Or they may have developed a rather secretive rebelliousness when young, aimed at not openly challenging their own parents, which remains as a mode of functioning throughout life. It is easier to deal with the difficulty at the time.

Summary

Adolescence is a time of great transformation. Parents, too, must change. Our teenagers need to feel secure within the home, to have a base from which to explore the world. They need to know they are loved and trusted as they venture forth to find a new identity for themselves. Much of their rebelliousness and defiance is linked to trying to separate from you, to look for their own way of being. This inevitably leads to conflict and hurt, both sides feeling misunderstood and unloved at times. For parents sometimes this growing up feels like a terrible loss – of their role, of their identity as well as of their little child. The distance between them can feel like a huge gulf. It is this struggle to be different, to be separate, which eventually gives the teenager the confidence and self-esteem to be strong and creative in the world and make positive relationships with others. It is also your blessing and encouragement of their freedom to grow up which makes them wish to be close to you.

5

Illness in the Family

> When sickness comes into the home, it does not merely take hold of a body, but weaves a dark fabric that entangles hearts and enshrouds hope. Sickness was like a spider spinning its thread around our plans, around our very breath, as day after day it swallowed up our lives.
>
> MURIEL BARBERY, *L'élégance du hérisson* (translated by Paul Haviland)

Families have always been affected by illness. Whether physical or mental, acute or chronic, inevitably it has dramatic repercussions. Within my particular framework of looking at boundaries and limits, I would like to show how illness can permeate the dynamic of the whole family, with its impact not limited to the ill person. It is hard to tell whether we are now more sensitive to these issues or whether certain kinds of illness are more widespread (asthma, eczema, food allergies, obesity, diabetes, etc). Children are facing increasing mental health problems, manifested by violence, depression and self-harm. Families affected by illnesses such as cancer and the loss of a young parent seem more prevalent. How do families maintain a sense of ordinariness when faced with extraordinary circumstances? Many daily routines that one could take for granted are questioned and the solutions are not obvious. My examples are specific and I have deliberately gone into detail to show just how dominant an illness can become in the family. The actual illness may differ, but the general impact of hospital visits, interventions, medication, grief, worry and terror will apply in some way to most families living with illness.

Babies

STAYING GROUNDED

Prematurity in the UK and other industrialized countries is the commonest cause of Infant Mortality.

Daniel was born at 28 weeks' gestation, 12 weeks premature. He suffered from Chronic Lung Disease and spent his first eight months in the Neonatal Intensive Care Unit, where I met the family. He started life on a ventilator to help him to breathe. Even when he returned home, he had nasal prongs which gave him a trickle of oxygen to facilitate his breathing. His parents were refugees and struggled with English. There were a few times while Daniel was in hospital when we thought he might die. After he returned home, he still had many hospital admissions and had to be re-ventilated on more than one occasion. Daniel's breathing eventually improved but he developed severe feeding difficulties, vomiting most of his feeds and not gaining sufficient weight. This led the paediatricians to consider carrying out a gastrostomy (a way to feed a baby who cannot suck or swallow sufficiently by inserting a tube straight into the stomach). When Daniel was eighteen months old, I was asked to join with the paediatric team in the hope that psychological help might prevent him having such an intrusive procedure.

In the beginning I tried offering a supportive setting to the parents, in which they could think about their own experiences with him and past losses and trauma. This was not very successful. However, at this stage Daniel's health was at such risk that I took on a more focused, active and directive approach. I was worried about possibly undermining his parents' confidence, but I was also acutely aware that a model of how to feed him might be helpful. Here is an extract of my meeting with them, where I suggested observing and giving him a feed, and considering with them what it might feel like to be Daniel.

Mr and Mrs Q speak about their anxiety over Daniel's poor weight gain. They are terrified he may never eat and will not survive. He is sitting on his mother's lap but quickly slips off to come and say hello to me. He clambers onto my lap and I encourage his mother to feed him. He gags immediately the spoon is offered, before even putting anything in his mouth and clearly finds it hard not to vomit. He gets off my lap and rushes around the room crawling everywhere. His mother offers him some food as he approaches her and he gags again and runs away. He goes to sit on his father's lap and seems more settled but still rather manic. He touches Mr Q's face and giggles. They play a bouncy, jolly, exciting game together. Mr Q gives Daniel a rusk which he sucks and then begins to cough violently. Mrs Q gets very alarmed and Mr Q starts patting Daniel's back quite hard. I feel they need talking through this difficult moment, to stop them becoming overwhelmed by his panic. I comment on how hard it must be for them at times like this, to remain calm and hopeful, when they have been through so much since Daniel's birth.

I am very aware that Daniel does not get a quiet time to try to begin eating. There is no boundary set to indicate the beginning of the meal. Daniel returns to my lap and again we attempt to feed him. This time, having observed his reactions, I am firm about keeping him still, enveloping him with my arms so that he wriggles less and feels held. His mother offers another spoonful. Daniel tastes it and starts gagging again. I soothe him with a hand stroking his back and my voice explaining that it is only food, tastes nice and that he can manage it. I believe it is crucial to hold him also with the musicality of my voice, to give it a calm and gentle tone. I am also quietly determined to stick it out and not let him go. I believe he can do this, as long as he has help. He starts to feed steadily, but his breathing becomes more and more laboured. I comment to his parents how hard he finds it to breathe and swallow at the same time. He carries on gagging at each mouthful and I carry on talking him through it. After a while he takes the food happily but shakes his head when it is in his mouth, as if expressing that he does and does not want it at the same time. I point this out to Mr and Mrs Q and that he needs an adult to make the food more appealing and to bear his ambivalence about it, without giving up, as well as giving him time to recover between each mouthful and re-stabilize his breathing. Mr and Mrs Q are surprised and say they've never seen him feed so well.

One important point in this encounter was that I gave a clear message to Daniel when he was going to eat. When he started to gag, on my lap, I believed it was possible to calm him down, give him time to breathe and stroke his back to help him to swallow. I tried to stay grounded and not let my fear turn to panic that he would choke and die. In other words, I tried to make the seemingly unbearable manageable by giving it meaning. I also had to resist the temptation, once he started to eat, to get carried away and want to get as much in as quickly as possible for fear he may stop at any moment. The prospect of the gastrostomy loomed large in my mind. I had to put a limit on myself and adapt to his pace. As a professional and not his mother, it was much easier for me to do all this than for his parents who had lived with the constant fear of his dying since his birth. They had lost any sense of hopefulness but also of appropriate limits for him. If he ran away from food then they felt helpless and allowed it, rather than tackling it together with him.

For the parents it is a huge task to detoxify what for Daniel is a painful and frightening experience both physically and emotionally. Not to be swayed by his flight from food and to remain firm and set containing limits is a Herculean undertaking. This is where others, professionals or family, can help. By being less flooded by the child's feelings they can model a different approach for both the parents and the child. They can keep hope and optimism about the outcome more alive than the understandably overwhelmed parent. By repeated and increasing positive experiences, both child and parent can modify their expectations and move closer to managing the task without assistance.

ATTENTION! ATTENTION!

George, a long-awaited baby, was born after the loss of a previous baby. He suffered a stroke during labour and has been carefully monitored ever since. He was diagnosed with very mild cerebral palsy but at risk of severe haemoplegia (paralysis down the right side of his body). When I first met George, at three months, he was a beautiful, bonny, blond baby. When watched closely, one could see that he under-used his right side, but it was not obvious. His loving and dedicated parents spent endless hours visiting hospitals for scans, check-ups, blood tests and physiotherapy appointments. George spent the vast part of his first year being attended to medically, at times as often as four times a week. He had sixty appointments in his first four months. Although his parents were told to treat him 'normally' they were also told to monitor him for a huge range of symptoms which could be cause for concern. They had to watch out for signs of neurological impairment and cognitive delay. They had to pay particular attention to the way he handled objects and ensure he used the affected side. They were trained by physiotherapists to dress him in a particular way and to manoeuvre his body to gain maximum benefit from everyday tasks. These early months, apart from being ridden with anxiety, made it nearly impossible to establish ordinary routines. Inevitably they were very attentive to his every squeak and movement.

As he has grown, thankfully beautifully and quite robust, they have become aware of how difficult he finds being set limits. In the early days, his mother particularly found it hard to let him cry at all. She was frightened that it might induce a fit. Also, because he had to put up with so many intrusive and difficult procedures, both parents wanted to spare him any additional discomfort. Saying 'no' often seemed almost unnecessarily cruel.

I have seen the family off and on since our first contact. George, now two and a half, still finds not being paid attention difficult, understandably as he has been so closely monitored all his life. From doctors, nurses, physiotherapists, grandparents to parents, everybody has been vigilant and closely engaged with him. Even though much of the attention has been quite unpleasant, he is used to receiving a lot of it. Working with him and his parents we have discussed how overblown his reactions to limits are, how quickly he gets upset if he is not attended to immediately, how hard he finds it to wait, etc. In observing him and thinking together we have been able to devise a more realistic approach to what does need attending to and what are the normal everyday frustrations any child his age must learn to cope with. Much of my work has been to try to hold on to George, the growing child, rather than George the baby who had a stroke and is at risk. Observing him with his parents has made it possible for us to notice together some of his strengths and also some of his own individual character. This is a welcome contrast to the need to observe him for signs of pathology. With time it has become easier to get a picture of George as a person and not just as a 'damaged baby'.

Feeling responsible for spotting any sign of delay or difficulty makes it very hard for parents ever to be 'off duty'. Recently George's mother told me how guilty she felt reading a book when he should be having his nap. She told him it was a quiet time and if he would not sleep then he should occupy himself in a calm and quiet way. He looked at her absolutely astounded, almost shocked that she could be with him in the room and pay attention to something else. However, after a brief moan and protest, he settled himself with picture books and stayed happily looking at them. This utterly sensible approach felt to her almost neglectful and she kept a wary eye on him while trying to read! Learning to be alone in the presence of someone else is a stepping stone to independence and the building of inner resources, so essential for healthy development.

Having someone reassure the parents that 100 per cent attention is not always helpful gives them the strength to resist the demand from their child. It is not selfish for the parent to want time for herself, and is good practice for the child to realize that people are not totally ruled by *his* wishes. For George it is crucial to learn this as he is now at nursery and having to share adults' attention with his peers. George's mother now realizes that you can't guard against the emotional impact of his early history. He had to go to appointments; he had to be tested and monitored. As the child grows stronger and the terror of losing him recedes, the parents grow in confidence and are more able to say 'no' to him. She says: 'It's doing as good a job as when going to all the appointments, just a different sort of responsibility.'

Two to Five

MAKING UP FOR LOST TIME

Difficulties resulting from the early months often manifest themselves later, when the child is a toddler.

Post-natal depression affects about 20 per cent of women in the first year after birth. It is widely recognized as the most common mental health problem affecting women following childbirth. For a long time the shame associated with feeling less than thrilled at having a baby kept many women from speaking about it. However, as its prevalence is now recognized, the subject is covered in the public sphere, through women's magazines and newspaper articles. By admitting to her own feelings of depression, Princess Diana (at the time very much regarded as a fairy-tale princess) showed that it could affect even those whose lives appear privileged and untouched by ordinary family problems. This helped the public to view post-natal depression within the context of an illness that can touch anybody and not as a personal failing. Privately though, it is still difficult to recognize in oneself and in others, and help is not always forthcoming. A survey carried out by Mind (a mental health charity) found that 'over two thirds of women had to wait a month or more for treatment, while one in ten had to wait over a year'. The majority of sufferers (75 per cent) were offered medication, while only a third was offered counselling (some therefore receiving both). Given the importance of the early months, it is critical that families get adequate and appropriate support in this crucial period.

However, depression can also occur slightly later: some mothers feel overwhelmed when babies start to toddle and their demands seem clearer, more relentless and exacting. For an older first-time mother, who is well established in her career and feels in control of her life, the arrival of a new baby can shatter her feelings of security and competence. For some there is a feeling of remoteness from the baby, of not bonding early. For others there is just the sense of feeling miserable and not enjoying the baby as they feel they ought to. The common thread is unhappiness at a time when people expect to be overjoyed.

Mrs B was referred by her psychiatrist because she had had severe post-natal depression following the birth of her first child (a son, Seth, now aged four) and was considering having a second child. I saw Mr and Mrs B for some time, both individually and as a couple, to discuss the various issues concerning them. We talked about Mrs B's fear of the depression returning, of the difficulties between them as a couple, as well as their expectations, wishes and dreams of what kind of family they wanted. As with many other young families, they had their first child at a period of Mr B's career when he had to work very long hours and was under constant pressure to prove himself within a new company. The baby in such circumstances is born in the context of a mother becoming quite isolated and of a father carrying the extra burden of responsibility of supporting an additional member of the family. This in many cases leads to friction and conflict within the couple – as it did with Mr and Mrs B – each feeling unsupported and at times misunderstood, in the face of the new challenges the baby brings. For many couples like Mr and Mrs B, the stress on their relationship can directly contribute to the mother's depression.

The aspect of the work I wish to focus on here is the impact Mrs B felt her depression had on her approach to rearing her son, Seth. Although the initial reason for referral concerned the prospect of a new pregnancy, what seemed uppermost in Mrs B's worries was how exhausting and demanding she found Seth. There was also a great feeling of guilt towards him for finding his childhood taxing. She was able to speak about this honestly and openly, although recounting how she felt was clearly painful. She was also worried that he was a difficult child and not very compliant at nursery.

I met Seth and found him to be a precociously articulate and able child. He was clearly bright and enjoyed showing off his skills and knowledge. In the room with him Mrs B was extraordinarily attentive and observant. She noticed and responded to his every word, question and activity, thoughtfully and with praise. He drew and wrote, even very long and complicated words (hippopotamus), and described his visit to the zoo in great and relevant detail. I imagined that, totally at odds with her own perception of herself, other mothers would be envious of Mrs B's success in bringing up such a bright and engaged little boy. However, it was clear that she felt 'on duty' all the time, that she carried an internal sense of responsibility for offering Seth the behaviour of an ideal mother, in contrast to how she felt both about herself and about him. I believe that, as she had found looking after him such hard work, she demanded 'perfection' of herself, something none of us could achieve, nor as we have seen in the rest of the book, is it desirable. Also, as she did not feel things came to her naturally, combined with her sense of guilt, it seemed she needed to make up for what she felt was her lack of 'love and warmth' by offering him a lot of devotion and stimulation. Seth, in turn, was accustomed to his mother's full attention and found it hard to wait. He would constantly call 'Mum, Mum, Mum', barely waiting to give her the chance to respond. We discussed how useful it would be for him to learn to wait his turn; in a conversation, for instance, she could ask him to say 'excuse me' if he wanted to speak to her when she was busy. Seth, of course, initially took this to be a preface to carry on speaking over her or whoever she was with. With time, however, he began to share her with others: a crucial skill for a young child about to start school.

Mrs B felt that being a good mother meant offering everything she possibly could. There was little indication that she felt she could take time to look after herself, let alone be entitled to do so, for instance, by asking Seth to entertain himself for a while or to be quiet or any of the other ordinary expectations one could have of a four year old. We spoke of Winnicott's idea of a 'good enough' mother (see p. 9). In contrast, a mother who is ever vigilant, always available and actively involved with her child ends up feeling resentful and exhausted. The child can become endlessly demanding and be seen at times as a tyrannical beast or demon. This is no help to either of them. Our work together consisted of 'allowing' Mrs B to claim back some of her time and to realize that this was not just better for her but also for Seth.

Some of the work with Mr B focused on untangling which emotions belonged to Seth and which to him. We saw that he was very identified with his son, wanting him never to feel any irritation or upset. This interfered with Mr B's capacity to support his wife in establishing useful boundaries for Seth. We were able to see more clearly that a degree of frustration and struggle would actually be helpful to Seth at this stage in his development. Both parents had felt they had to make up for time lost in the early months when the couple had been so stressed, with the result that Seth's moods and wishes overrode their own.

One can see with a situation like this – where a mother has been overwhelmed by having a baby and truly suffered through her depression – that bringing up a child is no easy matter. The guilt impedes an ordinary relationship. The feelings of being at the mercy of the child creates a dynamic where the child is dominant and fear and anger with him ensue. It was important, in addition to my meetings with Mrs B and with Seth, to enlist Mr B's participation in family life, both in supporting Mrs B but also in being firm with Seth and helping him to learn about limits and his place as a child in the family. There was also further work with Mr B but not of relevance to the point I am making here. I just want to highlight it so that the mother's depression is not seen in isolation; every member of a family contributes to its dynamics.

Mr and Mrs B decided to go ahead and have a second child. We carried on meeting through the second pregnancy and birth, and I am glad to say that life was much easier second time around!

In commenting on what was helpful about my intervention, Mrs B said, 'the revelation that saying "No" is not only okay but beneficial was liberating'. She went on to clarify that this was not for the obvious reasons, as 'everyone now knows that boundaries and limits are good for children'. What she felt was particularly useful was the thought that for healthy development, children need to experience their parents as 'not perfect'. She did not want to perpetuate in her son her own internal pressure to be perfect. She could see signs in him of being unduly upset if he could not achieve what he aimed for, if things didn't work out just as he wanted them to. This could lead to a rigidity and hinder his capacity to be adaptable. She did not wish him to grow up with that stress nor to experience it himself later as a parent and pass it on to his own future children.

A TOPSY-TURVY WORLD

In the UK, it is estimated that 1 in 100 children suffer from Autistic Spectrum Disorders (ASD).

Ahmed, a three year old, was diagnosed as having autistic features. He seemed to live in a world of his own, would sing to himself, dribble a lot, rock and would sometimes scream for lengthy periods for no apparent reason. His mother and father got so worried about him that they would do anything to stop him, initially complying with simple changes to their life but soon finding that they adapted so much to his ways that the whole family's life turned topsy-turvy. For instance, he would only eat pancakes and screamed if his mother turned the light on. She eventually found herself cooking at night practically in darkness. It was only when she spoke to the Special Needs Assistant at school that she realized how much her life was tailored to his fears and how far she had got from where she wanted her life to be.

In working with such children it is the norm to hear of parents' and siblings' lives being totally dominated by the child's difficulties. Such children may scream for hours, rock endlessly, bite themselves, repeatedly bang toys, etc. In public the child may behave in such a bizarre manner that the family feels ashamed and humiliated. Parents often say they will let the child 'do whatever he wants just so that he doesn't "start"'. They also frequently comply with rituals insisted upon by the child, for instance at bedtime, which may take hours, again in order to get some peace.

This may sound extreme but when a child seems genuinely distressed, parents tend to adapt their behaviour to what they feel he can tolerate. They may also believe that he simply does not understand what is asked of him, as if he were more like a wild animal and not quite human. It is also usually the case that they themselves live in a state of constant exhaustion and therefore can tolerate less than an outsider would. Unfortunately the message then given to the child is that he is right to be frightened and that they agree that it cannot be managed, rather than helping *him* adapt to what they think is acceptable. With time they become so used to walking on egg shells and anticipating crises that they demand less and less of the child; he therefore gets less practice at struggling with his difficulty. Often the impact of the screaming or tantrums on themselves and the siblings is what the parents, or family as a whole, want to minimize or soften. But here again, instead of helping the child adapt to the family, the family adapts to the child and everyone's life becomes shaped by the ill child's perception of the world. This kind of situation truly benefits from, if not to say requires, outside help, to support the family in making the difficult changes necessary for the child's development. Trudy Klauber, a child psychotherapist specializing in work with autistic children and their families, writes: 'Many parents become more consistent in setting firm limits and boundaries, particularly if they can really see that it has importance for the child's cognitive and emotional growth. They can be helped to normalize the abnormal and to have hope and expectations for the child.'

Primary Years

SPARING PAIN

Kitty, seven years old, has cystic fibrosis. She was born extremely ill and had two operations within her first three days, three blood transfusions in her first month and spent most of her first year in hospital. Her parents were told she might die and her first few months were touch and go. For a long time her parents had to monitor her very closely; they had to record her daily food intake, weight, urine and bowel movements. Although no longer in such detail, they still have to be vigilant regarding her weight and bowel movements. They were taught to do her physiotherapy very early, at one and a half months, so that she and they got used to it. At present Kitty is small for her age and continues to need occasional admissions to hospital. She has a daily intake of medication (15–20 tablets plus oral antibiotics and nebuliser) and daily physiotherapy, which is quite uncomfortable. This involves vigorous patting and vibrating to the chest area to loosen the sticky mucus which can clog up the lungs and digestive system. She has this twice a day for 15 to 40 minutes each session. The physiotherapy necessitates her staying still in various positions and this has been very hard at times, especially when she was younger and it was difficult to explain to her what they were doing and why. Just think of how difficult it can be to keep a toddler still for a nappy change, then imagine her lying still for physiotherapy which lasts this long! The patting is also loud and this too can be disconcerting. Her food still has to be carefully monitored and she tires easily. As she has grown stronger though, it is sometimes hard to believe she is so ill, because she looks so pretty and well. Generally she is a happy child, thriving at school and has many friends.

Although obviously her parents were very aware that her illness caused her emotional difficulties, she was referred to me primarily because she was having trouble settling to sleep at night. She would creep into their bedroom and would not settle in her own bed. She got up a few times through the night and therefore woke up tired in the mornings. This in itself worried them because of her vulnerability to infection. Although I worked with Kitty on her own, I just want to highlight some of the problems her parents had in setting her firm boundaries. As in the case of George, they carried huge anxiety about her survival due to her early infancy. Sleep is a particularly frightening time because for many of us, consciously or unconsciously, it is associated with death (see pp. 26 and 81). Also because Kitty had to put up with such painful procedures, operations and blood tests on a regular basis, her parents really wanted to spare her any pain. Additionally the daily physiotherapy made them feel guilty. Kitty's father also pointed out that the physiotherapy is considered a preventive measure, not a cure, and he was not so certain of its efficacy. Knowing a treatment is crucial helps but doesn't stop parents feeling they are inflicting pain or at best discomfort on their child.

Kitty was mostly compliant in her various treatments and created little fuss, and they admired her strength and were grateful to her for this. They wanted to make up for all the unpleasantness and distress her illness caused her. Letting go was very difficult for them, but they also enjoyed quiet, cosy times together, with no need for monitoring or medical intervention. So they found setting boundaries very difficult. Going to bed was the most obvious boundary, but there were minor ones throughout the day; most of the time they gave Kitty what she wanted and indulged her. They found talking about these difficulties with me helped to put boundaries into a more ordinary context. They could begin to think about what other people do and what is expected of a child of Kitty's age. This helped them to see her as a seven year old, as well as a child with cystic fibrosis. Although they knew this very clearly intellectually, emotionally it was difficult to resist their wish for lovely moments with her, not wanting to upset her or have a conflict, even if this meant not setting useful limits for her. They needed reminding that it was okay to upset her sometimes, in an ordinary way, just like any parent with a child who doesn't want to go to bed, pick up her clothes or switch off the TV and do some homework. When the need for limits is seen as necessary for a child's healthy emotional development, it becomes much easier to set boundaries.

With a child who requires treatment, medication or ongoing interventions for a chronic illness (diabetes as well, for instance) parents tend to want to 'spoil' her to counterbalance all the hardship she is going through. But the child can then end up demanding and less secure. Parents often feel unsure when the child really needs their help and when she could manage on her own (also very common in dealing with children who suffer from ME or Chronic Fatigue Syndrome). This is often a cause of conflict between the parents, who frequently hold different views. Sometimes they may feel that their good will is exploited. This can give rise to feelings of frustration or irritation within the parent, with themselves as much as with the child. It can also breed an unnecessary sense of dependency and helplessness in the child. In a paradoxical way boundaries are even more important for such a child, to build up her sense of agency and feeling in charge of herself. The illness can make her feel out of control and totally at the mercy of a fragile body. She may therefore try to control others so that she feels strong. Setting boundaries and expecting her to deal with some fears or worries by herself gives her the opportunity to practise something she finds hard to manage. In the long run, this will help build her confidence in a deeper way.

STOP THE WORLD I WANT TO GET OFF

A survey carried out in 2004 showed that 1 in 10 children in Great Britain aged 5–16 had a clinically recognizable mental disorder. It is now estimated that Attention Deficit Disorder/Attention Deficit Hyperactivity Disorder (ADD/ADHD) affects about 1 in 100 primary school children; Obsessive-Compulsive disorder (OCD) causes real difficulties and anxiety in functioning ordinarily in daily life. Studies put the rate of OCD at between 1–3 per cent of the general population in the UK. When there is severe impairment in functioning, medication is also used.

Peter was diagnosed with Obsessive Compulsive Disorder at the age of seven. He would panic and scream if things were not done his way. It could take hours to get him out of the house, making getting to school on time a daily ordeal. Bathing was a nightmare. He would inspect the bathroom and get extremely upset if he found a dead fly, for instance. Walking on the street he would hop around so as not to crush any bug which might be there. If his mother took him to a café and someone swatted a wasp, he would shout at them that they were murderers. In the face of his terror and his extreme reactions it was hard for his parents to carry out even simple tasks, such as getting dressed or eating breakfast. Taking a walk to the park was a major excursion and the idea of catching a bus to go further was unthinkable. He washed excessively and never felt clean enough, making his skin very red and sore. He refused to wipe his own bottom until he was ten years old, because he feared getting his hands dirty. His rituals took a long time and if they were interrupted or somehow changed he would have to start all over again. At night time his mother spent hours reassuring him that he and his loved ones would not die.

The question of ordinary boundaries did not even arise. As with the autistic child mentioned earlier, it becomes very hard to differentiate what the child can and cannot bear. The panic is contagious. As discussed in the first chapter, about babies, the parent can be flooded by the child's undiluted terror. It is very hard in those circumstances to distance oneself from the child's fear and to keep hold of reality. A crack in the pavement may be a huge dangerous chasm; a dead fly in the bathroom could be experienced as totally polluting, upsetting and filling the room with death. Nothing is ordinary. But how is the parent to know what is happening in the child's mind when he is just screaming and rigid with fear? It is not always clear what is upsetting him. Even if it is, who can control external reality so that there are never any cracks, never any damage? You cannot change the world; the child has to adapt to it. Parents also become very isolated; their universe becomes smaller and smaller. It is as if the child's anxieties take hold of them and their thinking and behaviour mirrors his fears. Their worries then escalate in a similar way to his. It is very taxing to organize to take the child out and the cycle begins: he may embarrass them, he cannot be left with other people, he may get hysterical and need to be removed physically from his obsessive repetitive behaviour or it would never end, etc . . .

In such circumstances, the parent who is mainly with the child, usually the mother, needs a lot of support to hold on to ordinary limits and to help her survive the strength of the child's feelings of terror, so that she can present him with a more reality-based reaction. Couples are often split in their attitude about how to deal with the child and frequently separate under the pressure, leaving one parent as the primary carer. At a time when the parent feels isolated and out of touch with ordinary life, it becomes extremely difficult to think straight and keep a sense of what she can expect and ask of her child. She may also feel unable to ask for help from friends and relatives. After a while parents can get used to living highly ritualized lives themselves to avoid any impending crisis. Eventually the parent may forget to offer what it is possible the child could tolerate, for instance asking them to be polite, to let them speak on the telephone, to wait for a minute and all sorts of small daily frustrations. I believe much effort is spent in helping such children but on the whole not enough is done to support the parents, even if only to encourage them to enlist help from around them. When the child is being seen by professionals and the parent is given strategies to work on with him, I believe it is too much to ask parents of such children to minimize the rituals or go through a programme with the child without supporting them extensively. The nature of the child's compulsions often appear unchangeable and parents need help in being guided at this time and given hope and strength to see difficult strategies through.

Adolescents

Depression, commonly thought only to apply to adults, is often undetected in children. It is thought that 1 in 10 teenagers (mostly female) could be affected by self-harm.

With the increased incidence of eating disorders and self-harm in adolescence, the issue of parents' boundaries resurfaces. In 'ordinary' adolescence young people are reaching out and forging their own place in the world. At times this may be very frightening and worrying for the parents but is part of a healthy developmental stage (see Chapter 4, p. 144). However, in some cases the young person's depression and self-destructiveness put them at serious risk and parents are needed and brought in to be more involved.

IT'S MY LIFE AND I'LL DIE IF I WANT TO

Helen is fourteen and bulimic. She keeps her over-eating and vomiting very secret and it is hard for her parents to know whether it is related to how she feels about her weight and body shape or whether it is more about being generally unhappy and depressed. The family has always been loving and close and this has come as a shock. They are very worried about her but she will not discuss any of this with them nor apparently with her friends. Therefore they feel they have no idea what triggers her bulimia. They are reluctant to challenge her on the whole range of usual parental complaints about teenagers: the mess she leaves around the house, treating the home like a hotel, her father as a credit card and her mother like a maid service, the endless telephone calls, and the lack of consideration for others. They do not want to upset her, in case it makes her ill. This leaves them feeling they are walking on egg shells all the time. They cannot tell what boundaries to set and how.

Helen's parents tried confronting her with their knowledge that she was making herself sick; they wanted to speak to her about it but she adamantly refused to acknowledge her bulimia and denied any unhappiness. After many attempts at getting closer to her, cajoling as well as getting upset with her, it became clear that she did not want them involved in her struggles. They eventually offered for her to see a psychotherapist, which she accepted. Her treatment is kept very confidential and her parents have little idea of whether Helen has even told her therapist of her bulimia. However, her mood is improving and she is moving somewhat closer to her parents. It is apparent that she is still being sick occasionally but she is far more open about her feelings. In this case it seems the young person needed to address her difficulties independently from her nearest and dearest and preferred to deal with a specialist.

Simon, fifteen years old, drinks heavily and had to go to hospital to have his stomach pumped. His parents also suspect he takes drugs. They sense he has a rather addictive personality and don't know how to intervene. As in the previous example, they have become frightened of doing anything that might upset him. They are reluctant to object to friends they think are not suitable because they worry this will only push him further into the hands of his delinquent peer group. They do not want to alienate him. They are unsure about how to speak to him for fear of seeming intrusive and controlling. Paradoxically, in their effort to prevent him putting himself at risk, they feel paralysed and unable to protect him. They find it difficult to think about what is appropriate. They fluctuate between wanting a rigid and total grounding and what they feel is a more thoughtful and sensitive approach. Unwittingly they shift from punitive to overprotective. Should they treat him as a much younger child who needs to be protected and supervised or can they risk treating him as a young teenager who could participate in his own recovery? Should they contact some of the other parents? Their anger against a system which does not protect children, allowing them to obtain drink and drugs, also interferes with their confronting him directly. Their belief in their capacity to parent is shattered and so at the time when they feel they most need to be parents, they feel least able to do so. They wish somebody else could tell them what to do.

There is no simple 'one size fits all' solution. We need to think about what suits each child within his family. In Simon's case it proved helpful for his parents to speak to him in a calm and loving way, communicating their concern for him. They were not judgemental about his friends nor did they express disappointment in him. They were firm about their responsibility as parents to protect him and decided not to allow him to go out at weekends for the following month. Instead they made an effort to spend time together as a family, doing things they all enjoyed. They managed this by communicating as a parental couple, being united in their approach and also discussing their difficulties with friends who had children of a similar age. Feeling less isolated and worried about being seen as 'bad parents' and realizing that many of their friends faced similar dilemmas helped them enormously. Their approach to Simon reinforced the message that there are ways other than being drunk to have a good time. It also helped Simon find his way 'home' when he had felt quite lost. In this case their solution made him feel cared for. Their stance gave him increased confidence so that when he started going out again he had less need to drink to excess. They also carried on involving

Simon in more joint activities and responsibilities (such as taking his younger sister out to the movies) within the family so that he had a secure base within the home as well as being able to be more adventurous with his friends.

Adolescence is the hardest stage for which to give practical advice, yet even with clear guidelines it is hard to be firm.

Melanie, sixteen, suffered from anorexia nervosa and after spending nine months in an eating disorders in-patient unit, she returned home. She was meant to keep some of the routines established in the unit, one of which was to eat within a given time frame, one hour for each of the three main meals, and half an hour for the three snacks. However, at home, she took far longer. The advice to parents was to stop the meal after one hour, whatever she had eaten by then. But her mother was not able to do this. When Mel pleaded for a bit more time, and then a bit more, the fear of her returning to hospital, the feeling that she desperately needed to eat, however long it took, the ultimate fear of her starving herself to death, interfered with her mother's capacity to carry out the advice given. Meals often took up to four hours and the whole day was mostly taken up with feeding Mel. The young person's terror, combined with the natural protective instinct of the mother, created a recipe for disaster in terms of boundaries. How can a mother refuse her child the most basic form of assistance with the most basic need and the one she provided first of all, food? But in this case the mother would have been more helpful to Mel in the long run if she had carried on with the hospital routine.

Eating disorders are one of the most difficult illnesses to shift and families feel paralysed. Here again the parent, usually the mother, finds it almost impossible to keep a sane hold on the world and a healthy perception of limits. In trying to make sure her child stays alive, the mother is not giving her daughter the opportunity of making that choice for herself, and then doing what is necessary; much easier said than done and a painful and frightening process.

The mother needs an objective person to help her let go and to help the young person take more responsibility for her own welfare. This could be a family member or friend but needs to be someone kind, consistent and strong.

My feeling is that as parents have become more liberal and accepting of almost everything their children do, it is very hard for adolescents to rebel. I believe it is one of the unhappy legacies of the 1960s generation. Where in the past teenagers could indulge (or pretend to) in 'sex, drugs and rock 'n' roll' as a rejection of their parents' morals, nowadays many children believe their parents have 'been there, done that, got the T-shirt'. I am not saying that this is true or that the past was better or that being understood by parents is bad per se. I do think this perception leaves them with ever more extreme options of how to demonstrate their rejection of adult values. The greatest hurt today's adolescents can inflict on their parents is to harm themselves. I worry that our over-protectiveness and our lack of strong limits pushes them to holding us to ransom, with themselves as hostages.

LOSING ONE'S OWN SENSE OF SELF

I would like to return to a different aspect of OCD – helping a parent deal with the changing needs of an adolescent as opposed to a young child.

When Peter was sixteen, he managed much of the daily routine more easily than when he was younger; he got himself to school with help and coped with full days. However, he was still dominated by his rituals and ruminating thoughts. He could not truly judge for himself when he was clean enough to come out of the bath, and his mother would end up nagging him to get out or the water would turn cold. He was often so disturbed by his thoughts that he could not get out of the house. He also paced up and down at night, unable to settle to sleep, plagued by his worries. My work with his single mother (the marriage had collapsed under the strain of his illness) began in his teenage years and focused on supporting her to start letting go a bit, to help her differentiate the tasks he really needed her for and those he could do for himself. Having lived with his OCD for so long she had lost sight of the fact that although he was still intensely ritualistic, he had also grown older. We always recalled the time I said 'His OCD doesn't mean he can't help you lay the table!' She had got so used to doing everything for him that she almost stopped expecting anything from him. She still treated him at times like a young child. I pushed her to try giving him more opportunities to deal with his difficulties himself or bear the disappointment of what his rituals denied him. For instance she would get very upset if he was slow to get ready in time to catch the school bus. She felt it was entirely her responsibility. I encouraged her to let him manage the time himself and see what would happen. He missed the bus occasionally and had to entertain himself during the day or make his own way to school. But both of them found that although difficult and upsetting this was not the end of the world. Some days were bad but others were manageable.

We also had to work on untangling her feelings from his and on my trying to help her get to sleep even if he wasn't able to. Her concern for him and her sadness at his torment haunted her nights too. Unfortunately, this then made home such a gloomy place for both of them; it became very hard to recharge their emotional batteries. She would be reluctant to go out with friends in the evening, partly through exhaustion but also because she felt so bad leaving him in what she knew was a bad state. Her sense of guilt could also stop her enjoying herself when she did go out. However, her starting to have a bit of a social life also gave Peter the sense that she would not always be at home and that he had to begin forging a social circle of his own. At a time of his life when his peer group spent most of their time together it was important that Peter did not always go home to his mother. Her absence could encourage him to venture out by himself more.

Much of my work was encouraging her to feel entitled to a good life. It was also difficult for her to keep hold of the lively, creative person she knew herself to be but felt she had lost. Many of her therapy sessions were spent asking about her, her feelings, wishes and dreams and reminding her that she too had a life rather than focusing on Peter. This is a common thread in families with illness; the carers lose sight of their own life, talents, desires, needs and aspirations.

Parents' Illness

When a parent is ill, the fabric of the household is disrupted.

CHANGING ROLES

Mrs C had a bad fall while horse riding. She was knocked unconscious and was hospitalized for an extensive period of time, far from home. At first there was a serious concern that she may be brain damaged. Fortunately she got help in time and was able eventually to make a full recovery. However, the period of her hospitalization and subsequent convalescence had a major impact on the family dynamics. Mr C was naturally preoccupied and worried. He had to juggle working hard to support the family as well as regularly visiting his wife and making arrangements for the children. Their eldest daughter Emily, fifteen, and their young son Joshua, nine, were often looked after by friends and relatives. At home, Emily took on the role of 'mummy' for much of the time, collecting Josh from after-school activities, keeping him occupied, ensuring he ate, brushed his teeth and got to bed. She also learned to cook simple meals and generally be in charge. Josh, due to his worry about his mother, became rather difficult, demanding and angry. Children often resort to anger when they are anxious. He strutted about the house like a little emperor. It was as if rather than being sad and frightened about his mother's state, he became outraged at the world for what had happened. He also became very controlling, which can be seen as a way of protecting himself from feeling the absence of his mother, who was usually in control of his daily life. For peace of mind and to manage more easily, Emily let him eat whenever and whatever he wanted, watch endless TV, etc. By the time Mr C came home or at weekends, he was too tired to enforce any rules or strict bedtimes. Both children were used to a very 'hands on' mother so the dramatic shift to having to look after themselves so much was difficult. Emily had a good model of how things should be done and fell into the part with some ease but at a cost. She identified with the role and almost forgot she was still a teenager. She saw far less of her friends, became very serious and a little bit arrogant. She would get frustrated and upset if Josh didn't pay her as much respect as he would to a parent. She became very close to her father, was pleased he could rely on her and enjoyed feeling she was like 'a little wife'.

When Mrs C came home, it was difficult to shift some of the dynamics back to a healthier and more age-appropriate stage with both children. Emily got offended if she was told off or if her mother wanted to run the household in a particular way. She felt patronized when asked about her plans or whereabouts, as she had got used to being her own boss. Josh on the other hand regressed dramatically and became very clingy and demanding, like a much younger child. Both parents had to work hard at reassuring them that their mother was alright now and that routines would be restored. Mr C had to be sensitive to his daughter's feeling that she had been demoted and help her to see that there was no longer any need for her to take on the extra responsibility and that in fact she would now be freer to lead her life. Joshua had to be reassured that his mother wasn't about to disappear again. The parents needed to acknowledge the impact of the accident and be patient and appreciative of the difficulties the children had been through. The family had a loving and strong foundation and a very supportive network and with this secure base they were able to re-establish the boundaries they had in place prior to the accident.

DEATH TAKES A GRIP

Mr X was in the last stages of dying with cancer. When I visited, the whole house was shrouded in darkness and silence, children hushed so as not to disturb or distress him. He had been very ill for some time and when entering the house it was difficult to feel that he was still alive; it was as if death had entered and taken a grip on everyone. Of course it was important to protect him but it was also important to allow those who were active to be so. The hustle and bustle of the children could have helped the father to remember that life still went on, and helped them all to hold on to life when faced with a death. Instead it was as if life was not allowed to show its face too boldly. This could be to spare him the disruptions of everyday life, being considerate of his feelings and doing everything in one's power to alleviate his suffering. But it could also be a manifestation of guilt at still being well, when a loved one is ill, or despair and disillusion with life itself, a giving up because it all seems so unfair or insurmountable. At times like this it is helpful for outsiders to breathe life into the home, not in a jolly, 'let's pretend nothing is wrong' way but in a strengthening way that helps the family believe life still has good things to offer, even if at times it is unbelievably painful. A limit needs to be set on how much the illness permeates the whole household.

LIFE IS A HEALER

In contrast Mr and Mrs K ensured that their two daughters (eight and five) held on to the business of life, as their mother also faced death. This was partly facilitated by the couple seeing myself and a colleague throughout the wife's illness. It provided the opportunity for her to voice her upset and anger. The couple at times argued vehemently and tearfully about his insistence on allowing the children to carry on having friends visit and on the energy he expended in thinking about them. Sometimes his wife felt she came last and she felt unprotected and unloved. But overall, as they discussed it, she too wanted her children to feel they could be themselves, even though at times it was overwhelmingly noisy and exhausting for her. As long as her feelings were acknowledged and she did not feel her illness was ignored, she was able to allow the spirit of life to permeate the house. The household was filled with the usual sounds of chatter, laughter and crying.

We were able to think about how she felt that life going on, as normally as possible, meant that nobody cared about her and also that in a way it seemed not to matter whether she lived or died. We could think about Mr K's need for life to go on ordinarily, partly as a denial of the painful loss ahead and possibly also expressing a sense of anger towards his wife for dying. Voicing these painful feelings allowed us all to think together. She became more able to ask for specific help (particularly noise levels being kept down) and they were able to talk together in a spirit of consideration and co-operation rather than antagonism and persecution. Mr K became more able to offer some of the simple acts she wished for, which he had been unable to do, for instance being held closely in his arms, his big frame making her feel more secure as she felt increasingly fragile. In discussing both their feelings in a safe setting we could see that what they were doing was trying to hold on to the goodness of life in spite of their grief, both for her benefit and especially for the children's. Mr K voiced that the therapy provided an ambience and a holding that enabled him to do what he felt he had to do.

The clear separation of the needs of the person who is ill and of those who are not helped the children and the family keep a sense of life going on, and thus gave everyone the permission to be themselves. One day, as Mrs K's condition worsened, Mr K told his eldest daughter that Mummy was now very ill and may die soon. She answered 'I feel very, very sad but is it alright if I feel something else as well?' A few days later, her school was opening a new dance studio and she was one of the children who was to perform in it. Again she was able to voice her dilemma, asking if it was okay to be excited about it. The children remained in and out of their parents' bedroom, even as Mrs K finally lay dying. It was not that she was being ignored at all, on the contrary she was not isolated in her dying and I imagine that if she was aware at the end, it was as she would have wanted it to be. I visited her on her final day, sitting by her side and having the children running in, sitting on the bed, talking to their Dad, asking him for mundane things, was extremely moving to me. It made death feel like a life process and far less terrifying than the way it is usually experienced. Many months after her death, in a session to review how he and the children were managing, Mr K said 'people say time is a great healer. That's nonsense. It is life which is the healer.'

My last two examples are rather extreme as they deal with parents who were dying, but the message would be even stronger for families where there is a chronic illness, which may last years. The important element is to try to bring as much ordinariness as possible into an extraordinary situation, rather than turning every day ordinary occurrences topsy-turvy to fit in with the unusual circumstance.

Summary

In the situations I have described in this chapter, the anticipation of what to expect of family life is shattered, as if the shared fantasy is a version of 'happily ever after'. Illness can be experienced as a bolt out of the blue. It shatters our life and sometimes our sense of ourselves too is fragmented. Our capacity to think clearly is undermined. It is as if, while we cannot understand what is happening, we lose confidence in understanding itself. We then act to ease pain in the moment, the here and now, rather than think about what is best for the whole family long term. I strongly believe that in these situations support needs to be given to whole families and not just the ill person. Where the illness dominates family life, it is almost impossible for the family to set appropriate limits without an enormous degree of help. More often than not, there is a need for professional input, intensive if possible, especially at the beginning of the illness. At this point difficulties are not entrenched and there is a greater degree of flexibility, as well as energy, in everyone.

When early infancy is fraught with difficulty and illness, it becomes very difficult for parents to differentiate between what is particular to the illness and what is developmentally appropriate. Holding on to usual boundaries can help normalize family life and also feed back to the child that there is a healthy and robust aspect to him, in the midst of all the worry. As the child grows older, it is important to keep a sense for oneself of common expectations and not give up on the child's capacity to change, albeit at a different pace. It is also crucial to try to keep as intact a sense of oneself, and not to turn one's life upside down to fit in with the ill person.

When the child enters school, hopefully there will be support from a greater network than the immediate family. Here too, I believe it is important for parents and siblings to try to maximize this contact with regular life and not close their own world up to fit in with the universe limited by the illness. Professionals can help the family to realize that this is healthy and helpful to all, and not something to feel guilty about. Similarly, it is important to allow others to look after the ill child and to trust that you are not the only one able or willing to manage. Friends and family are often willing to play a part and it is perfectly acceptable to consent to their offers. Trusting others with your child can be the greatest compliment.

Adolescence proves a more difficult time because it is when the child is expected to fly with his own wings. Parents are in the paradoxical position of needing or wanting to protect their child as well as needing and wanting to help him take responsibility for his own welfare. Again, external professional help is often required.

When an adult is ill or dying they may need to work on putting boundaries on themselves. They may have to struggle with their own helplessness which makes them feel childlike and needy. They may need assistance in understanding and untangling what makes them feel bad. Allowing others to live and thrive will in fact bring them far more support and love when they need it. It will also go a long way to leaving the survivors less guilt ridden.

Afterword: Couples

> We need, in love, to practise only this: letting the other go. For holding on comes easily; we do not need to learn it.
> RAINER MARIA RILKE, *'Requiem for a Friend'*

As I conclude this book, it would be comforting to think that once we have thought carefully about the issue of saying no, with special emphasis on us as parents, we have come to the end of the story. However, it will be obvious to you by now that the whole business is about processes, about on-going relationships. We are therefore confronted by similar feelings and concerns in all spheres of our lives – in our marriages, partnerships, with our own parents, and with colleagues at work.

I have briefly referred to the difficulties couples have in agreeing about their children. It is hard at times to say no to the other, to state our point of view yet negotiate a common solution. We may just each deal with a problem individually, to avoid conflict, even when we know a united front is helpful. There are many occasions when we face this struggle, and not all of them are related to being the parents of our children. I wish to have a very brief glance at some of the themes I have covered but in relation to our relationship as a couple, without the children.

Yes as a Gift

Most partners wish to please the other, to offer something unique and special. One of the ways we do this is by agreeing with them, by supporting them, by being positive about their endeavours. We speak of our 'other half', of being a 'unit', we praise and envy those who are never apart, who are joined, bound, who are soul-mates. In this portrayal of the ideal couple, there is little acknowledgement of the value of saying no, of being different. Just as with children growing up, there is no doubt that in a couple a great degree of fitting in, of togetherness, gives pleasure, boosts attachment and fosters growth. It provides a secure base. However, I strongly believe that here, too, there is a need for a space, a gap between people, for them to be able truly to develop. A convolvulus, although closely wrapped up around another plant, does not support it, but rather twists it and occasionally stifles it.

Always saying yes, even if you really feel you totally agree with your partner, may make him and you feel there is no difference between you. This can be a comfort at times, but leaves life very static; there will be little movement. If there is a change in one or other partner, it can then feel like a terrible let-down, a breaching of an unspoken pact. At times this has catastrophic consequences and the couple breaks up.

The other pitfall is saying yes because you want to please, even when you would rather not agree. You may be happy to offer your partner something which gives him pleasure – your 'yes' is a gift. However, if this is your regular pattern, you may end up feeling resentful and taken for granted. You are then grudging and irritated and may blame the other: 'I do so much for him/her, and never get any recognition!' In fact you were the one who chose this avenue; you have responsibility for your actions.

You may also worry about causing displeasure and of the consequences for both of you. These may just be related to emotions – you may not want to feel mean, selfish, unkind. You may not want to see someone you love disappointed, irritated or angry. So you avoid the confrontation. In fact saying no can be immensely liberating for both partners. It encourages divergent points of view, it offers the opportunity for change. If you each state your view, you open up the possibility of finding a joint agreement which respects both people's individuality, and which is arrived at together, not by one person making assumptions about the other. You also discover that you may disagree and still be close. By saying no yourself, you give permission for the other also to say no.

Hall of Mirrors

We have seen that we get a picture of ourselves through the reflection in other people's eyes. This is particularly the case with couples; the image may make us feel like the most wonderful creature on earth or it may cause us great distress. Many of us will be upset by how we are perceived by our partner, especially when this does not match our image of ourselves. 'He always makes me sound like a nag', 'She thinks I work all hours for the fun of it' are frequent complaints.

If we consistently get a particular kind of feedback, we may start to doubt our own sense of ourselves. So a wife who believes she is stating a differing point of view, but feels she is answered as if she has attacked her husband, may desist and retreat. Similarly, a man who is feeling vulnerable and left out and then feels treated like a spoiled child, may in turn react in a childish manner. It is very painful to see yourself depreciated in another's eyes and shakes your self-confidence. It is hard to hold on to your own self-worth and easier to slip into believing the picture you see. As adults, when faced with an image we do not like, we will turn inwards to see if it matches other experiences from our past. This will make us more or less resilient in the face of hurt feelings. We will have a greater or lesser capacity to say 'no' to a portrayal of ourselves which we feel does not fit.

Another distortion which has an impact on how we relate to each other is what sometimes happens during absence. Just as our partner has an image of us in mind, so do we. This can evoke strong feelings while we are not together.

Diane and Henry would often argue and both felt unappreciated by the other. Although they always made a point of eating their dinner

together after their baby went to bed, they often dreaded what this quiet time might bring up. As the time approached for Henry to come home, both would imagine the mood of the other, anticipating a problem. Diane, having spent all day at home with the baby, was looking forward to seeing Henry and chatting about what she felt was the 'outside world', but she imagined Henry would be exhausted from work and want to flop in front of the TV. She expected him to be cut off from her and self-centred. In turn Henry, tired from his day, anticipated that Diane would be miserable and depressed from being at home and was already irritated with her.

At the point at which they then met, they were both carrying very specific, negative pictures of the other in their minds. When we discussed how often this evening time proved difficult, we found that the 'person in mind' was the one each responded to at the end of the day, regardless of what mood the other one was actually in. The fantasy of the other replaced reality. When they got together, there was no way that they could see that the other was actually very different from the one they had imagined. For instance, Henry might really have been looking forward to a nice meal and talk with his wife. Similarly, Diane might have had a lovely day and felt refreshed and enthusiastic about a peaceful evening at home with him. The work with this couple involved them becoming aware of the gap between who they expected to see at the end of the day and who was actually there. They had to suspend judgement and wait openly to check out the other's reality. Each had to take responsibility for what he did to his partner in his mind, i.e. turn him into something other than he is. We all do this to an extent.

GHOSTS

One of the ways we can understand this distortion is by looking at what comes between us and a clear view of our partner.

Raj works tremendously hard. He has always been an achiever and has very high standards. His father was a very exacting man whom Raj felt was never satisfied. He also thought his father expected him to achieve all that he himself had not managed. This had been a heavy burden, particularly in adolescence. Now in his marriage Raj frequently becomes very defensive when he experiences his wife making any request. It throws him back to his childhood and he feels put upon, as if all his hard work is not valued and that whatever he does, more is required.

Fiona went to a boarding school where she was severely bullied. Although she timidly indicated to her parents what was going on, they seemed not to hear or chose not to do anything about it. She became a rather withdrawn and shy adult, not speaking up for herself. Although happily married, she finds it difficult to disagree with Simon, her husband, and to object to any of his requests openly. He is frequently aware that she is upset and that it is to do with him, but does not know what it is he did. During disagreements, Fiona is in touch with her memories of being bullied and tends to cower and give in. She then reacts to Simon as if he were her violent bullying schoolmates, and not the husband who is merely disagreeing with her.

We can see how the ghosts in the nursery, discussed in Chapter 1, follow us throughout life. At various times, feelings from the past are reawakened by people in the present. For instance, in the earlier example, Henry came from a family where his mother was frequently ill and depressed. Whenever Diane was a bit miserable, it made him feel terribly oppressed, to a greater degree than her mood could account for. It was as if her unhappiness turned her in his eyes into the down-trodden, depressed mother of his childhood memory. In this way, his picture of Diane became distorted and he reacted to more than just her behaviour.

Similarly, the distorting mirror of the past altered Raj and Fiona's images of their spouses. Raj reacted as he did as a young teenager, resentful of the pressure put upon him. Fiona cowered and retreated as she had done as a child. They were all transferring on to their partners an image from their experience with others, from their childhood families. This, in turn, changed the way the partners reacted to them, tapping into their own histories. So couples often enter a confusing and circular pattern which is hard to break, partly because the interfering factors are not conscious, and so are not available for thought.

This is a problem seen repeatedly in psychotherapeutic work with couples. It is very difficult for either partner entering this atmosphere to say something akin to, 'I don't know who you were thinking of when you spoke to me like that, but it did not feel as if it had anything to do with me.' Just as when the young child sometimes feels his mother turns into the Wicked Witch of the West and she needs to remind herself that in truth she is not, so in couples sometimes we have to remind ourselves of who we actually are. We need to be able to say no to being turned into something other than what is true to us.

Together but Separate

Negotiating our differences is a struggle at all ages. The capacity to hold on to our own feelings, not to be invaded by those of the other, is essential if we are to have a relationship which is based on mutuality. We need to be sure enough of ourselves to say, 'No, this is not about me,' to the picture presented of us, if we feel it does not fit. We need to be confident in holding our position of difference. We should be able to say, 'No, I don't really want to do that', if that is how we feel. The whole issue of being separate, of being individual, has been the subject of my book. One way we demonstrate our separateness is by saying no to others.

Couples have to negotiate differences all the time. We often want to push the other to conform to an ideal, to fuse with us in terms of expectation. We can see that our own histories play a part in how we view the other and ourselves. We also repeat to some extent patterns of relating which are familiar to us. We try to get others to see things as we do. But in order to be truly reciprocal, an adult relationship has to involve two separate people who choose to be together. We talk of ties that bind us, yet what makes for real closeness is freedom of choice. We need to say no to our urge to impose our way, to hold the other so tightly either to ourselves or to our image of them. In order to be close, we have to let go. We will then be able to engage in an equal and true exchange.

References

Epigraph

xii Bhownagary, F., 'To My Children', *Poems*, 1997, Calcutta: Writers Workshop, p.17.

Preface to Second Edition

xi National Family and Parenting Institute Survey, IpsosMori, August and September, 2006. Seventy-two per cent of parents have watched at least one parenting programme; 83 per cent who have watched report finding a technique from the programmes helpful. When asked which source of help they found most useful, TV programmes scored 33 per cent (friends and family, 59 per cent; schools and playgroups, 40 per cent).

xii Christakis, D. A. et al. (2004) 'Early Television Exposure and Subsequent Attentional Problems in Children', *Pediatrics*, Vol. 113, No. 4, April 2004, pp. 708–13.

Introduction

xvii Phillips, A. *On Kissing, Tickling and Being Bored*, 1993, London: Faber and Faber, p.xvi.

xviii Shah, I. *The Exploits of the Incomparable Mulla Nasrudin*, 1973, London: Pan Books Ltd, p.26.

1 Babies

1 Tagore, R., from 'The Beginning' in *The Crescent Moon*, in *Collected Poems and Plays of Rabrindanath Tagore*, 1982, London: Macmillan, p. 57.

1 These studies are described in Brazelton, T. B. and Cramer, B., *The Earliest Relationship*, 1991, London: Karnac Books.

3 Bion, W. R., *Learning from Experience*, 1962, London: Heinemann.

7 Stern, D., *The First Relationship*, 1977, Cambridge, Massachusetts: Harvard University Press.

8 Murray, L. and Cooper, P. J. (eds), *Postpartum Depression and Child Development*, 1997, New York and London: The Guildford Press.

9 Winnicott, D. W., 'good enough' mothering is an idea coined by Winnicott and recurs throughout his writing.

11 Winnicott, D. W., 'Transitional objects and transitional phenomena' in *Through Paediatrics to Psychoanalysis*, Collected Papers, 1992, London: Karnac Books and the Institute of Psychoanalysis, p. 238.

26 Hood, T., *The Death Bed*.

31 Klein, M., 'Weaning' in *Love, Guilt and Reparation and other works, 1921–1945*, 1975, London: Hogarth Press and the Institute of Psychoanalysis.

34 Brazelton, T. B., *Touchpoints*, 1993, Addison-Wesley Publishing Company, p. 233.

36 Fraiberg, S. *Ghosts in the Nursery: a psychoanalytic approach to the problems of impaired infant-mother relationships*, in *Clinical Studies in Infant Mental Health – the First Year of Life*, 1980, London and New York: Tavistock Publications, p. 164.

41 Brazelton, op. cit., p. 79.

42 Brazelton, op. cit., p. 369.

2 Two to Five

46 Tagore, R., from 'When and Why' in *The Crescent Moon*, in *Collected Poems and Plays of Rabrindanath Tagore*, 1982, London: Macmillan, p. 58.

47 Bettelheim, B., *The Uses of Enchantment*, 1978, Harmondsworth: Penguin Books.

63 Harris, M., *Thinking about Infants and Young Children*, 1975, Perthshire: Clunie Press, p. 45.

75 Sendak, M., *Where the Wild Things Are*, London: Bodley Head, 1967; Picture Lions, 1992.

76 *Young Minds Magazine*, October 1996, London, p. 12.

80 Daws, D., *Through the Night*, 1989, London: Free Association Books.

81 Tennyson, A., 'In Memoriam A.H.H.' in *The Works of Alfred Tennyson*.

81 Tomlinson, J., *The Owl who was Afraid of the Dark*, 1984, Harmondsworth: Puffin Books.
83 Herbert, Frank, *Dune*, 1975, London: New English Library, p. 14.
87 Saint, Exupéry, A. de., *The Little Prince*, 1998, Mammoth, p. 64.
88 Ibid., p. 68.
94 Ibid., p. 66.

3 The Primary School Years

95 Dahl, R., *Matilda*, 1989, London: Puffin Books, p. 69.
99 Holmes, E., 'Educational Intervention for Pre-School Children in Day or Residential Care', in *Therapeutic Education*, Volume 8, no.2, Autumn 1980, p. 9.
106 Bettelheim, B., *The Uses of Enchantment*, 1978, Harmondsworth: Penguin Books, p. 7.
112 Britton, R., 'The Oedipus Situation and the Depressive Position', in Anderson, R. (ed.), *Clinical Lectures on Klein and Bion*, 1992, London and New York: Tavistock/Routledge, p. 40.
124 Phillips, A., *On Kissing, Tickling and Being Bored*, 1993, London: Faber and Faber, p. 72.

4 Adolescents

144 Takeko, K., in *The Burning Heart – Women Poets of Japan*, translated and edited by Rexroth, R. and Atsumi, I., 1977, New York: The Seabury Press. p. 69.
163 Waddell, M., *Understanding your 12–14 year old*, 1994, London: Rosendale Press, p. 69.
164 Winnicott, D. W., 'Transitional objects and transitional phenomena', in *Through Paediatrics to Psychoanalysis*, Collected Papers, 1992, London: Karnac Books and the Institute of Psychoanalysis.
165 Winnicott, D. W., 'Adolescence – struggling through the doldrums', in *The Family and Individual Development*, 1965, London: Tavistock Publications, p. 85.
169 Shakespeare, W., *Much Ado About Nothing*, Act II, Scene III
170 Dolto, F. and Dolto-Tolitch, C., *Paroles pour adolescents – Le complexe du homard*, 1989, Paris: Hatier.
179 Sondheim, S., 'Jet Song' from *West Side Story*.
185 Robertson, J. and J., *Separation and the Very Young*, 1989, London: Free Association Books.

5 Illness in the Family

187 Barbery, M., *L'élégance du hérisson,* 2006, Paris: Éditons Gallimard.

187 European Commission Research website: Respiratory allergies such as hay fever and asthma are increasing on a global scale. In Western Europe, the prevalence of asthma is reported to have doubled in the last ten years, while allergies in general are also on the increase. One striking feature is the geographical variation in allergy prevalence in Europe – major international studies focusing on asthma and allergies in children have found a twenty-fold difference between various European centres, the reasons for which remain to be explained.

187 National Eczema Society website: 'The prevalence of eczema has been on the increase for the last 30 years. Recent research conducted at Bristol University has demonstrated that one in three children from 0–42 months have eczema. This dramatic, though not unexpected, increase in prevalence is matched by other allergic conditions.'

187 US Department of Health and Human Services website: 'The prevalence of food allergy is growing and probably will continue to grow along with all allergic diseases,' says Robert A. Wood, M.D., director of the pediatric allergy clinic at Johns Hopkins Medical Institutions in Baltimore.

187 Overweight and obesity are increasing rapidly. In England, the percentage of adults aged 16–64 who were obese has increased by over 50 per cent in the last decade. High levels of overweight and obesity among children are likely to exacerbate the trend towards overweight and obesity in the adult population, since compared to thin children, obese children have a high risk of becoming overweight adults. Between 1995 and 2004 the prevalence of obesity almost doubled among English boys (from 11 per cent to 19 per cent) and increased by over a half in girls (from 12 per cent to 19 per cent).

187 British Heart Foundation Statistics website: The prevalence of diabetes is increasing. Since 1991, the prevalence of diagnosed diabetes has more than doubled in men and increased by 80 per cent in women. Extrapolating from the Health Survey for England data it is suggested that the prevalence of diagnosed diabetes in the UK will be about 7 per cent for adult men and 5 per cent for adult women by the year 2010. This would equate to approximately three million people with diagnosed diabetes.

187 Breast cancer incidence in women has increased from one in twenty in 1960 to one in eight today. National Breast Cancer Foundation, 'Signs and Symptoms', 2006 and American Cancer Society, 'Overview: Breast Cancer', 2006.

188 Grenville, F. Fox (2002) *Fetal and Maternal Medicine Review 13*: 195–211, Cambridge University Press.

193 Priest et al., 2003, 'Out of the blue? Motherhood and depression' in *Mind Week Report,* May 2006.

197 National Autistic Society website referring to research by Gillian Baird published in the Lancet, 2006.

199 Klauber, T., 'The significance of trauma and other factors in work with parents of children with autism' in *Autism and Personality* (eds Alvarez, A. and Reid, S.), 1999, London: Routledge.

199 BUPA website, 2007: Cystic fibrosis (CF) is an inherited condition. It affects organs in the body, particularly the lungs and digestive system, which become clogged with sticky mucus, making it difficult to breathe and digest food.

202 National Statistics online, 2004. Survey of the mental health of children and young people in Great Britain.

202 NHS NICE Clinical Guideline 31, 2005. Obsessive-compulsive disorder.

204 NICE Guideline Development Group, Professor Peter Fonagy, Professor of Psychoanalysis and Chair: 'Depression in children is more common than many people realise and often goes unrecognised. Around 1 per cent of children and 3 per cent of adolescents will suffer from depression in any one year. It can severely impact on school performance, self esteem and making and retaining friendships. It can lead to a greatly increased risk of mental health problems in adult life and at its most serious it can dramatically increase the lifetime risk of suicide, from 1.3 per cent in the general population to 6 per cent.'

204 Royal College of Psychiatrists website, 2005.

Afterword: Couples

217 Rilke, R. M., 'Requiem for a Friend' in *Selected Works of R. M. Rilke* (ed. Mitchell, S.), 1982, London: Picador, p. 73–87.

Index

PACC
MOYA • 07736836088
Lifemodel